COBALT

Edited by **Khan Maaz**

Cobalt
http://dx.doi.org/10.5772/intechopen.68255
Edited by Khan Maaz

Contributors

Hasan Eskalen, Süleyman Kerli, Şükrü Özgan, Yasemin Yıldız, Philip Greenwood, Kw Guo, Theophilos Ioannides, Maria Smyrnioti, Petrică Vizureanu, Mirabela Georgiana Minciuna, Gianina Iovan, Simona Stoleriu, A O Oyetunji

Published by InTech
Janeza Trdine 9, 51000 Rijeka, Croatia

Publishing Process Manager Iva Simcic
Technical Editor SPi Global
Cover InTech Design team

First published December, 2017
Printed in Croatia
Legal deposit, Croatia: National and University Library in Zagreb

Additional hard copies can be obtained from orders@intechopen.com

Cobalt, Edited by Khan Maaz
p. cm.
Print ISBN 978-953-51-3667-5
Online ISBN 978-953-51-3668-2

For EU product safety concerns:
IN TECH d.o.o., Prolaz Marije Krucifikse Kozulić 3, 51000 Rijeka, Croatia,
info@intechopen.com or visit our website at intechopen.com.

Meet the editor

Dr. Maaz Khan is working as a senior researcher in PINSTECH, Pakistan. He received his PhD degree from Quaid-i-Azam University and postdoctorates from South Korea and China. He is working on fabrication of nanomaterials by different chemical and electrochemical techniques and studying their structural, magnetic, optical and electrical characterizations. Dr. Khan is the author of more than 75 articles in peer-reviewed journals and also the author and editor of many books in the field of materials science. He is working as *editor-in-chief* of Journal of Materials, Processing and Design and *executive editor* of International Journal of Nano Studies and Technology and is also serving as an editorial board member of few more journals.

Contents

Preface

The book is composed of seven chapters that describe the synthesis, characterization and application of cobalt in recent technologies. Various topics are discussed in the book, such as: the general aspects of cobalt chemistry, synthesis of cobalt and cobalt oxide by different chemical techniques, substitution and redox properties of some organo-isocyanide cobalt complexes and effect of cobalt on morphology of microhole formed by micropunching and a review of radionuclide cobalt-60 as a fine-sediment tracer. Each chapter is written clearly and precisely and illustrates easier understanding of the phenomena described in it. This book addresses not only scientists in specialized fields, but also doctorate students, postdoctoral researchers and technical professionals.

All authors that contributed to this book are very active and competent in their field, and therefore this book provides up-to-date information about the concerned topics. The content of the book provides fundamental preparation needed for further study of advanced topics on cobalt and its application. At the end of each chapter, proper references are included that can lead the readers to the best sources in the literature and help them to go into more details about the concerned topic.

I am grateful to all the authors who are experts in their fields for helping me to complete this project and also to the entire InTech publishing team for making this project possible. I am very thankful to the Publishing Process Manager Ms. Iva Simcic for her cooperative attitude during the entire reviewing and publishing processes. I hope that this book will provide an opportunity to the readers to strengthen their knowledge and capabilities in the relevant field.

Dr. Khan Maaz
Pakistan Institute of Nuclear Science and Technology
Pakistan

Chapter 1

General Aspects of the Cobalt Chemistry

Yasemin Yildiz

Additional information is available at the end of the chapter

http://dx.doi.org/10.5772/intechopen.71089

Abstract

This chapter aims to collect and summarize the chemical properties of cobalt and some new cobalt compounds. It deals with the progress of cobalt chemistry. Cobalt has been substantial in both chemical reactions and within many compounds. Some of them are heterocyclic reactions, cobalt-based catalyst and cobalamin. Also, it discusses variety of applications of cobalt in a wide range of areas and toxicity of cobalt. The studies carried out in this area so far have enabled and will be continued to be responsible for producing unknown and difficult reactions. This survey of the recent literature illustrates the fact that many different approaches on cobalt and new cobalt compounds are being used in many different areas.

Keywords: cobalt compounds, cobalamin, synthesis, catalyst, areas of usage

1. Introduction

The word cobalt is derived from the sixteenth century German term kobold. Glass, glazes and blue dyes for pottery were the oldest known uses of cobalt. For instance, cobalt compounds were used to dyeing pottery by Egyptians and Babylonians in 1450BC. Cobalt was set apart from copper ore by Swedish chemist Georg Brandt in 1735. After 1900, a new corrosion resistant alloy was invented, which is referred to as Stellite. Furthermore, aluminum-nickel-cobalt ($AlNiCo$) magnets caused an important evolution in 1940. They were used as the first electromagnet [1].

By 2000, scientists had carried out lots of experiments related to the synthesis of new cobalt compounds, examination of antibacterial properties of cobalt compounds and cobalt-catalysed reactions. For example, Kumar and Garg synthesized cobalt(II) complexes of tetradentate Schiff bases of the type [2].

Nowadays, cobalt and cobalt compounds possess a variety of applications from industry to medicine. This is because of its unique properties such as a high-melting point (1493°C) and retaining its strength to a higher temperature, being ferromagnetic with high thermostability and multivalent. Cobalt is one of the abundant metals in the Earth. Global reserves of cobalt are approximately around 7 million tons [3].

This chapter aims to collect and summarize the chemical properties of cobalt and some new cobalt compounds. The studies carried out in this area so far have enabled and will be continued to be responsible for producing unknown and difficult reactions. This survey of the recent literature illustrates the fact that many different approaches on cobalt and new cobalt compounds are being used in many different areas.

In order to briefly discuss about cobalt chemistry in this section, we begin with cobalt electron configuration. Cobalt has the chemical configuration $[Ar]4s^2 3d^7$ and has oxidation states Co(II) and Co(III). Other properties of cobalt are summarized in **Table 1** [4].

Cobalt has some inorganic compounds and complexes. Some chemical and physical properties of cobalt and several inorganic cobalt compounds are mentioned in the following statements.

Dilute sulphuric or hydrochloric acid dissolves slowly, but nitric acid rapidly dissolves and produces the cobalt-II ions [5]

$$3Co + 8HNO_3 \rightarrow 3Co\left(NO_3\right)_2 + 2NO + 4H_2O \tag{1}$$

Cobalt gives cobalt(II) chloride by dissolving in dilute hydrochloric acid [4].

$$Co + 2HCl \rightarrow CoCl_2 + H_2 \tag{2}$$

It is insoluble because it acquires passivity in concentrated nitrate and sulphuric acid [6].

Another oxidation number of cobalt is +3. But this ion can only be found in the complex. Bare cobalt releases oxygen from acidic and neutral environment. All cobalt oxides dissolve in hydrochloric acid [5].

Cobalt generally forms cobalt(II) and cobalt(III) compounds, but there are cobalt compounds, which have $4+, 1+, 0, 1-$ oxidation states [4]. It is more stable than those compounds which

Atomic number	27	Atomic mass	58.93
Electron Distribution	$[Ar]3d^7\ 4s^2$	Oxidation Number	+2,+3
Melting Point	1495°C	Boiling Point	2870°C
Electronegativity	1.8	Isotope	59
Density	8.9 g/cm^3	Atomic Volume	6.7
Ionization Energy	757.6 kJ/mol		

Table 1. Other properties of cobalt.

consist of cobalt-III ions. In general, cobalt-II compounds dissolve in water; however, CoS, $CoCO_3$, $Co(OH)_2$, $Co(CN)_2$ *and* CoC_2O_4 are insolvables. Cobalt-II is quite a little hydrolysis. Both cobalt ions do not have amphoteric properties [5].

It is known that all common cobalt compounds have octahedrally coordinated to nitrogen or oxygen ions, all of which include three cobalt ions, two of which are 3+ and one of which is 2+, similar to the title compound.

$Co_3(1-Hdatrz).6(H_2O)_6.\ (NO_3)_8.4H_2O.(Hdartz-3,5-diamino-1,2,4-triazole$ [7], was the most attractive of these that include three cobalt ions. Two of these ions are 3+ and one of these ions was 2+ [8].

Cobalt is found together with iron, copper, nickel, silver and arsenic in nature. Cobalt closely resembles nickel so that they are being identified as "twins." Although cobalt is used in practice to plate iron, doing alloy is the most significant practice of cobalt. Important cobalt alloys and its uses are summarized in **Table 2**.

The most known minerals are linnaeite, Co_3S_4 *or* $(CoNi)_3S_4$; smaltite, $CoAs_2$ *or* $(CoNi)As_2$ and cobaltite, $CoAsS$. Cobalt oxides are utilized in ceramics as smalt, a cobalt potassium silicate [9]. Important cobalt alloys and its uses are shown in **Table 2** [10].

Cobalt compounds which give blue color to glasses can also be made radioactive and can be used for treatment of some cancers. Further, it is used for the treatment of deep-rooted tumors. This shows that radioactive cobalt can be used as a tracer [11].

It is commonly known that salts and complexes of cobalt are catalysts for the selective oxidation of alkanes and selective epoxidation of alkenes. Scientists have performed several experiments on these catalysts. Several of them are discussed in this chapter [12].

In order to make liquid fuel, cobalt catalyst is used in the Fischer-Tropsch process. Also, cobalt and molybdenum are utilized as a catalyst in hydrodesulphurization of petroleum. These processes make use of refining of liquid fuels [13].

Nickel-cobalt-boride ($Ni-Co-B$) catalysts are used to design an efficient hydrogen generator for portable fuel cell applications [14].

The chemistry of cobalt complexes has attracted a lot of attention in recent years on account of their applications, among others, in biological systems such as antimicrobial agents and antibacterial agents.

Alloy name	Percentage composition	Features	Places used
Stellite	50–60 Co, 30 Cr,20 W,Mn, C	Hard and pourable	Cutting and drilling tools and mould
Vitallium	65 Co, 25 Cr, 5Mo	Resistant to wear and tear	Gas turbine wings
Magnet steel	35–60 Co, 10–25 Ni,Fe	Can be magnetized	Magnet

Table 2. Important cobalt alloys and uses.

To design novel drugs, medicinal chemistry has benefited from the properties of metal ions. Hence, this has caused to have clinical application of chemotherapeutic agents for cancer treatment, such as cisplatin [15].

Some of these works were mentioned as follows. In 1952, the first biological activity of cobalt compounds was acquainted where cobalt(III) compounds of bidentate mustard move as if it were hypoxia selective agents [16].

Bauer and Drinkard prepared and identified several new cobalt(III) complex compounds of easily oxidized ligands. These were salts of cobalt amines, inner charge complexes, and a variety of cobaltates. The structural formula was $Na_3[Co(CO)_3 3H_2O$ which was the simplest formula [17].

It is commonly known that salts and complexes of cobalt are catalysts for the selective oxidation of alkanes and selective epoxidation of alkenes. Scientists have carried out lots of examination on these catalysts. Several of them are *cobalt (III) acetylacetonate (acac)/O$_2$e, cobalt (II) – (salen complexes/O$_2$/isobutyrylaldehyde,cobalt (II)calix,pyrrole complexes/2 – ethylbutyraldehyde/O$_2$,Schiff base cobalt complexes/aliphatic aldehydes or b – ketoesters, polymer supported cobalt (II)/O$_2$,H$_2$O$_2$, (N – hydroxyl – phthalimide/Co(OAc)$_2$/O$_2$,cobalt (II)phthalocyanine* [18, 19].

Cobalt studies are still continued due to the fact that they have a wide variety of functions and many applications, especially in pharmaceutical technology.

2. Used places

Cobalt has many applications in a wide range of areas. A solution of *cobalt (II)chloride* can be utilized to measure moisture in the air. When the weather is humidity, paper is pink in color. Also, it may be used as an invisible ink [11]. Cobalt II salts are red in the case of complex-ion, that is, solvated with water, and it is dark blue when dehydrated. Invisible ink is made using this feature. If the diluted $CoCl_2.6H_2O$ solution is written on a sheet of paper, nothing can be seen. If the paper is wetted, the blue patch can be read [6]. This color change helps find that gel depletes the amount of time as dryer.

It is commonly known that salts and complexes of cobalt are catalysts for the selective oxidation of alkanes and selective epoxidation of alkenes. Scientists have carried out lot of on these catalysts. Several of them are *cobalt (III) acetylacetonate (acac)/O$_2$, cobalt (II) – (salen complexes/O$_2$/isobutyrylaldehyde,cobalt (II)calix,pyrrole complexes/2 – ethylbutyraldehyde/O$_2$,Schiff base cobalt complexes/aliphatic aldehydes or b – ketoesters,polymer supported cobalt (II)/O$_2$,H$_2$O$_2$,(N – hydroxyl – phthalimide/Co(OAc)$_2$/O$_2$,cobalt (II)phthalocyanine* [18, 19].

Exciting results have been obtained from these studies over the last two decades. For example, it was found that there was redox activity of quinone ligands and potential for forming compounds. These formed compounds may involve in a number of electronic states due to the combined electrochemical activity of the cobalt ion and one or more quinone ligands [12].

Cobalt compounds can be made use of treatment of some cancers [11].

Magnetic, wear-resistant, and high-strength alloys are chiefly made by cobalt metal. It is suitable for applications such as desulphurization of hydrocarbons, the removal of nitrous oxide, and the emerging technology of converting natural gas to liquid hydrocarbons due to its unique catalytic properties. Because of allowing manufacture of highly effective cutting tools, it is also suitable for base industry application. It is used in both gas turbines and aggressive working setting owing to high-temperature resistance, hardness and wear characteristics of cobalt while alloyed with other metals. It may be also contributed to improve the operating efficiency by means of durability and wear capacity [3].

Besides, cobalt is generated a significant part of vitamin B12 in biochemistry. Vitamin B12 is a *tetra aza macrocyclic ligand complex of Co*. Therefore, it is a vital mineral for survival. Inorganic form of cobalt is matter that nourishing for bacteria, algae and fungi in chemistry. In addition to all these applications, cobalt is used as a catalyst in some reactions. For instance, cobalt-based catalyst is essential in reactions involving heterocyclic synthesis. Cobalt carboxylates are ordinary catalysts which also utilized in paints, varnishes, and binding resin.

Cobalt catalysts are practiced in many reactions, which are the synthesis of heterocycles. One of them was cobalt-catalyzed oxidative free-radical cyclization of alkyl bromides in 1986 [20].

Another example of these reactions is $3, bis-\{2,2,2-trifluoromethyl-1-[(1R,4S)]-1,7,7-trimethyl-2-(oxo-O)bicycle\,[2.2.1]hept-3-ylidene]ethanolate-O\}-cobalt\,(II)$ that Hartung et al. found.

In order to make liquid fuel, cobalt catalyst is used in the Fischer-Tropsch process. Also, cobalt and molybdenum are utilized as a catalyst in hydrodesulphurization of petroleum. The process made use of refining of liquid fuels [13].

In biochemistry, Vitamin B12 ranked sixth coordination position of Co(III) that is included cyanide ion and so it has another name, cyanocobalamin. Cobalamin complex is joined to $5'deoxy\ adinosl$ unit through the C atom of the CH_2 group that is called Vitamin B12 coenzyme. Both of them are diamagnetic.

Vitamin B12 is indispensable for advanced creatures, but it is not vital for plants. It can only be synthesized by certain bacteria. The living creatures obtain it through the food. Vitamin B12 is digested in the ileum by a mucopolysaccharide, which is present in the gastric juice. If vitamin B12 is lacking or not digested, various diseases come into play. One of them is pernicious anemia. There is no absorption of cobalamin in patients with parenteral anemia. This is shown that it is vital for synthesis of hemoglobin. Besides, all of these indicate the biochemical significance of cobalt [21].

3. Some common cobalt compounds

Cobalt II salts are red in the case of complex-ion, that is, solvated with water, and it is dark blue when dehydrated. Invisible ink is made using this feature. If the diluted $CoCl_2.6H_2O$ solution is written on a sheet of paper, nothing can be seen. If the paper is wetted, the blue color patch can be read [6].

Besides, it is used to say how much moisture present in air. When there is humidity in weather, paper becomes pink [11].

3.1. Cobalt II sulfate, $CoSO_4.7H_2O$

It is dark red that gives double salts with alkali sulfates and is a form of monoclinic prisms [6].

3.2. Cobalt nitrate, $Co(NO_3).6H_2O$

It is separated from water in the form of red monochlorine crystals. If $Co(NO_3)$ is strongly heated it will form cobalt-2,3-oxide (Co_3O_4). With careful heating, Co-3 oxide is obtained [6].

$$2Co(NO_3)_2 \rightarrow Co_2O_3 + 4NO_2 + 1/2O_2 \tag{3}$$

$Co(NO_3)$ wield in treating the certain types of cancer due to it can be made radioctive [11].

3.3. Cobalt-III-sulphate, $Co_2(SO_4)_3.18H_2O$

It is obtained by anodic oxidation of $cobalt-II-sulphate$ and it gives a pair of salts which are opposed to the alum with potassium sulfate. Salts of cobalt-III in aqueous solution are easily converted into salts of cobalt II. On the contrary, salts of cobalt II also readily constitute salts of cobalt-III [6].

4. Synthesized new cobalt compounds

Cobalt was investigated by many scientists over years. These works ended up with many cobalt compounds. For example, Kumar and Garg synthesized cobalt(II) complexes of tetradentate Schiff bases of the type CoL. $[H_2l = C_2OH_{16}N_2O_2.$ (H_2dsp), $H_{18}N_2O_2.$ (H_2dst), $C_{20}H_{15}N_3O_4(H_2ndsp)$ and $C_{16}H_{16}N_2O_2(H_2salen)$. They found that $C_{20}H_{14}N_2O_2Co$ complex has the minimum and $C_{16}H_{14}N_2O_2Co$ complex has the maximum activation energy. They found that all the complexes were crystalline, brown-colored substances, non-hygroscopic and insoluble in non-polar solvents unlikely partially soluble in DMF and DMSO. Also, they evaluated complexes by UV-visible, IR, and magnetic studies and figured out various thermodynamic parameters using TG/DTA [2].

The chemistry of cobalt complexes has attracted a lot of attention in recent years on account of their applications, among others, in biological systems such as antimicrobial agents and antibacterial agents (DNA studies and cytotoxicity studies) [22].

Some of this works were mentioned as follows. In 1952, the first biological activity of cobalt compounds was acquainted where cobalt(III) compounds of bidentate mustard move as if it were hypoxia selective agents. Then, some compounds demonstrated significant activity against bacterial strains and against leukemia and lymphoma cell lines [20]. Afterward, lots of studies were carried out on anti-microbial, anti-fungal and anti-oxidant activities of cobalt compounds [16].

By 1960, several new cobalt(III) complex compounds of easily oxidized ligands were prepared and identified. These were salts of cobalt amines, inner charge complexes and a variety of cobaltates. The structural formula was $Na_3[(Co(CO]))_3]$. $3H_2O$ which was the simplest formula [17].

It was reported that *cobalt (II)phthalocyanine* (*CoPc*) and its derivatives show catalytic activity for many reactions. They compared the catalytic activities of the *CoPc* species with those of the other first row transition metal phthalocyanine complexes. They were also shown that CoPc-modified glassy carbon electrodes catalyze the oxidation of *o − cresol,m − cresol,p − cresol,4 − chlorophenol*, 2 − *chlorophenol* and *pheno*. Coating with *CoPc* species increased stability of GCE (cobalt(II) phthalocyanine-modified glassy carbon electrode) [23].

It was demonstrated that a complex form of cobalt(III) was an active catalyst for the selective oxidation of alkylaromatics using air. The air was used as the source of oxygen in the absence of solvent [24].

Park et al. reported the synthesis of "solid solution" and "core-shell" types of well-defined *Co − Pt* nanoalloys smaller than 10 nm. The formation of these alloys was carried out by redox transmetalation reactions. $Co_2(CO)_8$ and *Pt(hfac)hexafluoroacetylacetlacetonate* were generated in the formation of "solid solution" type alloys such as $CoPt_3$ nanoparticles (**Figure 1**). Obtained nanoparticles were moderately monodispersed. Park et al. firstly reported the usage of redox transmetalation reactions for the synthesis of two different types (solid solution and core-shell). Various types of nanoalloys can be synthesized with this kind of redox transmetalation strategy. It is possible to fulfill next generation requirements with magnetic alloys, because of monodispersed particle size of less than 10 nm scales [25].

When pure metals like Fe, Co and Ni and their metal alloys utilized in magnetism, it is difficult to use them because of their oxidation in air. Moreover, this difficulty increases when the particle size gets smaller. So, the stability of particle enhances with a variety of methods. One of them is deposition of insulating shells on the nanoparticles surface. A procedure that leads to air-stable Co nanoparticles was carried out by Gedanken et al. [26]. They pretended that the formation of a carbon shell on the nanoparticle surface increases stability. But, acquired particles were not uniform [27].

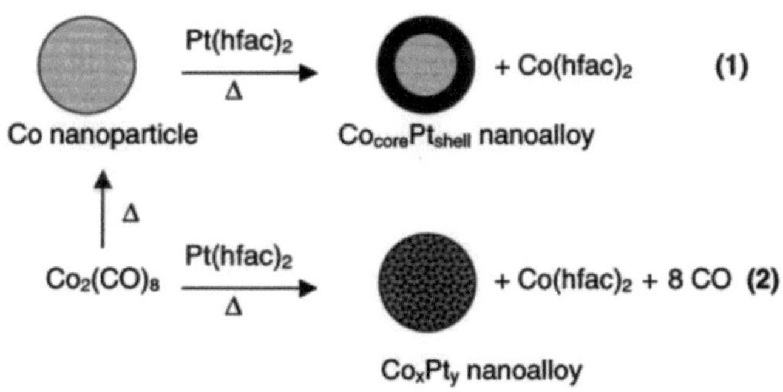

Figure 1. Synthetic routes of core-shell and solid solution type nanoalloys via transmetalation reaction.

Kobayashi et al. informed that it was a procedure, which allows the preparation of Co nanoparticles of various sizes in aqueous solution and their coating with well-defined silica shells. They found an easy chemical method for the synthesis and stabilization of magnetic and amorphous Co nanoparticles. These Co nanoparticles were surrounded by homogeneous shells of silica. This novel type of composite magnetic nanoparticles has potential applications, both in the field of ferrofluids and in magnetic storage media. The controlled assembly of $CoSiO_2$ nanoparticles is currently being studied by Kobayashi et al. [27].

The formation mechanism of the tubular structure of Co_3O_4 was studied by Li et al. and they presented a hypothesis. They prepared Co_3O_4 nanotubes by calcining electrodeposited cobalt nanowires embedded in an anodic alumina template (AAT). Oxidation, phase transition, evaporation and recrystallization processes were once believed that it had played a key role in the formation of Co_3O_4 nanotubes. Scientifically, it was a noteworthy case and these obtained nanotubes also find a wide range of applications in industry [28].

In vitro antimicrobial activity of cobalt(II) complexes was studied. These ligands were $[CoL_2Cl_2]$, where $L_1 = aminobenzimidazole$, $L_2 = 1 - benzyl - 2 - aminobenzimidazole$, $L_3 = 1 - (4 - methylbenzyl) - 2 - aminobenzimidazole$. All of them and their cobalt(II) complexes were examined in vitro antimicrobial activity against *Pseudomonas aeruginosa*, *Bacillus* sp., *Staphylococcus aureus*, *Sarcina lutea* and *Saccharomyces cerevisiae*. They found that none of the compounds were importantly impressive against yeast *Saccharomyces cerevisiae*, but 2-aminobenzimidazole complex was efficient, which moderately restrained the growth of yeast. Similarly, all ligands and their complexes were evaluated for activities one by one and they found that in case of benzimidazole, nucleus had 4-methylbenzyl group at the N1 atom, the antimicrobial activity increased. They concluded that the substituted ligands and cobalt may get a role in the antimicrobial activity [29].

It was reported that synthesis of cobalt nanoparticles using a polymeric microfluidic reactor by the reduction of $CoCl_2$ in tetrahydrofuran (THF) using lithium *hydrotriethylborate* $(LiBH(C_2H_5)_3$ as a reducing agent and $3 - (N, N - dimethyldodecylammonia)propanesulfonate$ (SB_{12}) as a stabilizer. They investigated experimental conditions such as flow rates, growth time and quenching procedure. They found that Co nanoparticles with mainly fcc structures were composed with a high kinetic energy level (e.g., high-flow rate). Co nanoparticles with mainly hcp structures were preferred when it was be conditions with a low kinetic energy level (e.g., low-flow rate) and short-growth time. But, the crystal structure of Co nanoparticles switched to the metastable phase, while Co nanoparticles generated at the low kinetic level grew for a longer time. This submits that by controlling the reaction kinetics, acquired nanoparticles with different structures can utilize as potential application in microfluidic reactors [30].

Ingersoll et al. carried out nickel-cobalt-boride $(Ni - Co - B)$ catalysts by a chemical reduction method and their catalytic hydrolysis reaction with alkaline $NaBH_4$ solution. They aimed to design an efficient hydrogen generator for portable fuel cell applications. The reaction temperature in the hydrolysis of alkaline $NaBH_4$ solution, a function of the concentrations of $NaOH$ and $NaBH_4$ were evaluated in their work. The rate of hydrogen generation was surveyed using $(Ni - Co - B)$ catalyst. The rate of hydrogen generation was increased with lower $NaOH$ concentrations in the alkaline $NaBH_4$ solution and decreased after reaching a maximum value at 15 wt.% of $NaOH$ [14].

By 2008, cobalt nanoparticles using thermal decomposition of [*bis*(*salicylidene*)*cobalt* (*II*)] in the presence of *TPP* and *oleylamine was studied*. The obtained cobalt particles were stable in hydrocarbon solvents to air oxidation and had cubic cobalt crystallinity. The hysteresis loops of the obtained samples were found to have soft magnetic behavior, enhanced coercivity (Hc) and decreased saturation magnetization (Ms) in comparison to bulk materials. After having contact with air, the hysteresis loop of Co nanoparticles also revealed ferromagnetic property saturation magnetization (Ms) and coercivity (Hc) values of about 56.2 emu/g [31].

Bruijnincx and Sadler investigated the design of modernist metal-based anticancer drugs that include recent literature. Many novel chances for anti-cancer drugs were gained low systemic toxicity and ability of coming from the top of the drug resistance. In their work was been presented with different examples of promising offer. These studies caused expansion toolbox of medical inorganic chemistry [15].

It had been studied with three species of bacteria, in order to found the activity of the schiff base and their complexes. In this work, it was found best results with diameters (30 mm), due to the inclusion of Co(II) ion. This was shown that the complexes have more antibacterial activities than the free Schiff bases [32].

Kumar and Chandra synthesized cobalt compounds that exhibit significant antifungal activity [33].

Pannu et al. synthesized $[CoLCl_2]_2$ and $[Co(L)_2(H_2O)_4][CoLBr_3]_2 2H_2O$. They found that latter compound has magnetic response [34].

The research was carried out in cobalt catalysis and they obtained cost-effective catalysts and milder conditions for existing C–H functionalization. Their studies also paved the way for unrivaled chemical transformations. They found two answers for two questions in this study. First, low-valent cobalt catalysts may imitate the reactivity of noble transition metal catalysts to C–H activation. Second, in C − H functionalization, cobalt catalysts may present matchless reactivity and selectivity and these properties provide a way for undergoing mysterious and hard synthesis transformations, up to now. Given examples were branched-selective hydroarylation of styrenes, ortho-alkylation of aryl imines with secondary alkyl halides and migratory arylzincation of alkynes. Their work brought out novel difficulties and possibilities. It was necessary to fully understand the reaction mechanism and the nature of the catalyst due to further growth in this area [35].

Gaëlle et al. synthesized and investigated two complexes, a cobalt(II) complex $[Co(phen)_3(NO_3)_2]2H_2O$ (1) and a novel Co(III) complex with mixed ligand [*diazido bis*(1, 10 − *phenanthroline* − *k2N, N'*)*cobalt* (*III*)]*nitrate*$[Co(phen)_2(N_3)_2]NO_3$(2). They examined their antimicrobial activities in vitro against eight pathogens (four bacteria and four fungal species). **Figure 2** showed the histogram of MIC against bacterial species. The results showed that the complexes were very active; however, complex 1 is more active toward the bacteria, but complex 2 is more active toward the fungi. The MICs values for complexes are summarized in **Table 3**.

Both of them may stand for decent nominees as an antibacterial (1) and antifungal agent (2). However, Gaëlle et al. continued the relevant works [17].

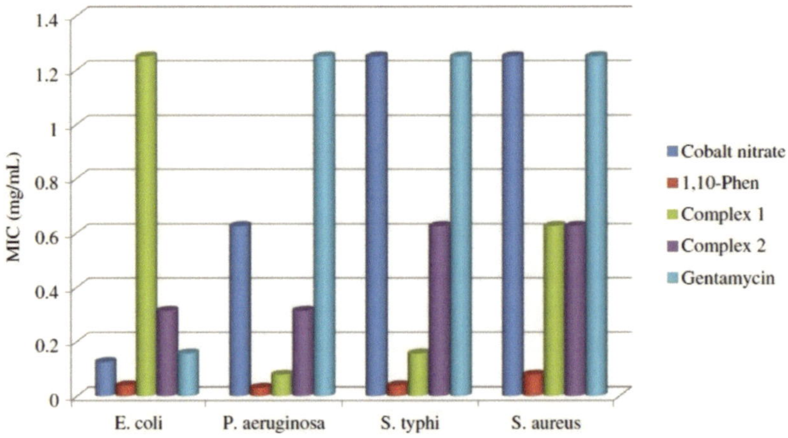

Figure 2. Histogram of MIC against bacteria species.

Three new cobalt(II) coordination compounds were reported (i.e., $Co(HL_1)Cl$ (1), $Co(HL_2)Cl$ (2) and $[Co(HL_3)Cl\ (CH_3)_2CHOH]$ (3) by Morcelli et al. [36]. Structures of complexes (2) and (3) involved the naphthyl group and contained α and β-naphthyl groups, respectively. Also, complexes (2) and (3) possess more activity than complex (1). The results of this work showed that there was effect of the isomerism on the antitumour activity. Moreover, complexes (2) and (3) paved the way for the death of cancer cells without affecting the normal cells. In this chapter, it is significant to determine the kind of the structure of the ligand and the kind of metal center on cytotoxicity [36].

	Species	**MIC(mg/mL)**						
		$Co(NO_3)_26H_20$	**O-Phen**	**N_3^-**	**Complex 1**	**Complex 2**	**Gentamycin**	**Nystatin**
Bacteria	*E. coli*	0.125	0.039	1.156	1.25	0.313	0.156	>2.5
	P. aeruginosa	0.625	0.039	0.078	0.078	0.313	1.25	>2.5
	S. typhi	1.25	0.039	0.156	0.156	0.625	1.25	>2.5
	S. aureus	1.25	0.078	0.625	0.625	0.625	1.25	>2.5
Fungi	*C. albicans* ATCC 12C	0.625	0.039	1.25	0.156	1.25	>2.5	>2.5
	C. albicans ATCC P37037	0.625	0.078	0.625	0.156	0.313	>2.5	>2.5
	C. albicans ATCC P37039	0.625	0.156	1.25	0.156	1.25	>2.5	>2.5
	C. neoformans	1.25	0.039	2.5	0.078	1.25	>2.5	>2.5

Table 3. MIC (mg/mL) of the complexes.

In 2016, three homospin compounds of chain structures that were linked to pentagonal bipyramidal Co(II) units were synthesized (**Figure 3**). Moreover, this work demonstrated the preparation of higher dimensional coordination polymers [37].

Cody et al. synthesized two new metal thiophosphate anions, $[Co(Co(P_3S_8)_2)_2]^{4-}$ and $[Cr(P_3S_9)_2]^{3-}$. These new compounds indicated that it is the applicability of ionothermal synthesis to two new metals in the thiophosphate system. Both syntheses demonstrate the versatility of ionothermal synthesis for preparing new thiophosphate compounds [8].

Two new mono- and dinuclear Co(II) complexes namely $Co(tdmpp)Cl_2]_2.2H_2O$ (1) and $[Co_2(tdmpp)Cl_4]$ (2) (where $tdmpp\,\frac{1}{4}1,1,3,3-tetrakis\,(3,5-dimethyl-1-pyrazolyl)propane\big)$ were prepared by Azizolla Beheshti et al. [38]. This work showed that bacterial inhibition ability of these compounds increases from tdmpp to the dinuclear complex 2 (**Figure 4**). Furthermore, it is demonstrated that the metal complexes are more active than free ligands [38].

Also, cobalt is used as a catalyst in some reactions. For instance, cobalt-based catalysts are essential in reactions involving heterocycle synthesis. Adam et al. carried out catalytic hydrogenation of nitriles to primary amines. They used $Co(acac)_3$in combination with *tris* [2 (*dicyclo hexylphosphino)ethyl*]*phosphine* as a catalyst. It catalyzes the selective hydrogenation of nitriles to give the corresponding amines [39].

The $[Co(2-hydroxy-1-naphthaldehyde_2(DMF)_2]$ complex 1 was synthesized by Zeinab Asgharpour et al. and successfully supported on modified Fe_3O_4nanoparticles using *tetraethylorthosilicate* (*TEOS*) and $(3-aminopropyl)trimethoxysilane$ (*APTMS*). Moreover, it was stated as $Fe_3O_4SiO_2APTMS$ *complex* nanocatalyst (**Figure 5**). It was a useful heterogeneous catalysis system that there were properties of solid catalyst candidates as $Fe_3O_4SiO_2APTMS$ *complex* 1 such as easy preparation, mild reaction conditions, high yield, ease of catalyst separation and recyclability [12].

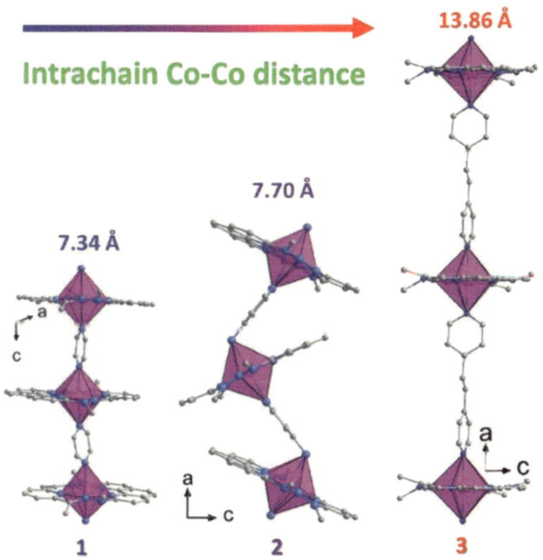

Figure 3. The 1D chain structures of complexes **1–3**. Hydrogen atoms are omitted for clarity.

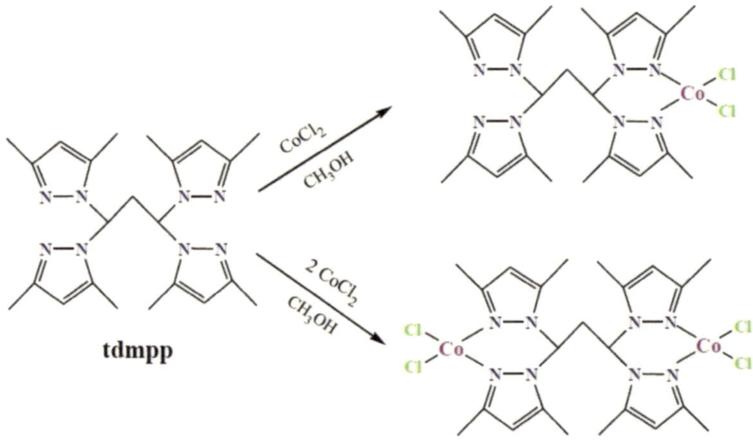

Figure 4. Synthesis procedure for compounds **1** and **2**.

Ko et al. carried out the preparation of hollow cobalt oxide and cobalt selenide microspheres and their Na-ion storage properties. Hollow cobalt selenide microspheres were classed with the hollow cobalt oxide microspheres. Hence, it was applied as an anode material for NIBs. Selenide microspheres showed high initial discharge capacity and high initial Coulombic efficiency as well as good cycling and rate performances for Na-ion storage. Also, they had structures that improve electrochemical properties by means of optimizing the electrolyte system used for Na-ion storage. On the other hand, hollow cobalt oxide microspheres were favorable from the point of their high initial capacity and low voltages for Na-ion storage as anode material for NIBs. They had leveragable cycling and rate performances. So, both of these microspheres were act as a promising anode material for NIBs [40].

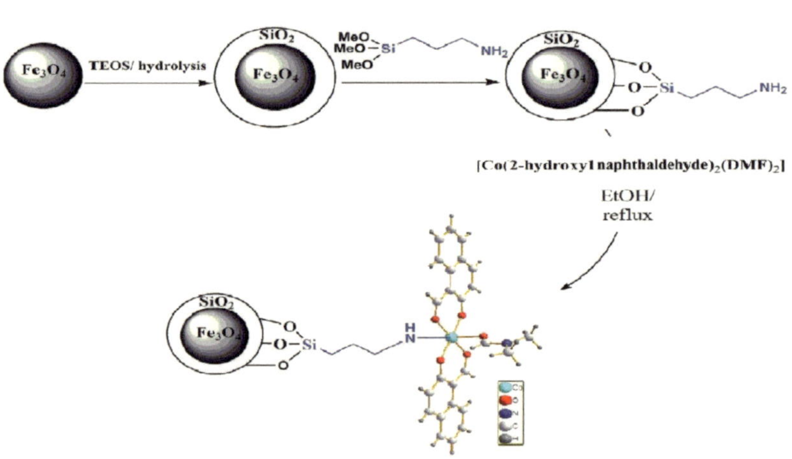

Figure 5. Preparation steps of Fe_3O_4@SiO_2@APTMS@complex 1.

http://dx.doi.org/10.5772/intechopen.71089

Mondal et al. prepared mixed *NiCoMn − Oxide* nanorods, which was a new class substance for supercapacitor application. They carried out process for synthesis of mixed *NiCoMn − Oxide* through a facile hydrothermal technique and their electrochemical capacitive properties. Structures were characterized by XRD and SEM. A variety of measurement showed that there was potential to utilize as electrodes for supercapacitor devices in this substance. Moreover, it was considered that there were cost-effective methods for extraction of raw materials. Therefore, mixed oxide was preferred as raw materials for large-scale applications such as supercapacitor electrodes [41].

It was made with fabrication of Ni NWs decorated with Co NPs by using two-step etching and deposition technique in 2016. This study showed that nanoparticles dispersed on the surface of nanoparticles as shown in **Figure 6**. They declared in this study that this technique has been used for the first time in order to manufacture nickel nanowires which are adorned with cobalt nanoparticles [42].

Montazerozohori et al. synthesized a new nanostructured cobalt(II) bromide complex with a bidentate Schiff base ligand. Particles' size of complex in nanodimension size by XRD and SEM analyses is shown in **Figures 7** and **8** [43].

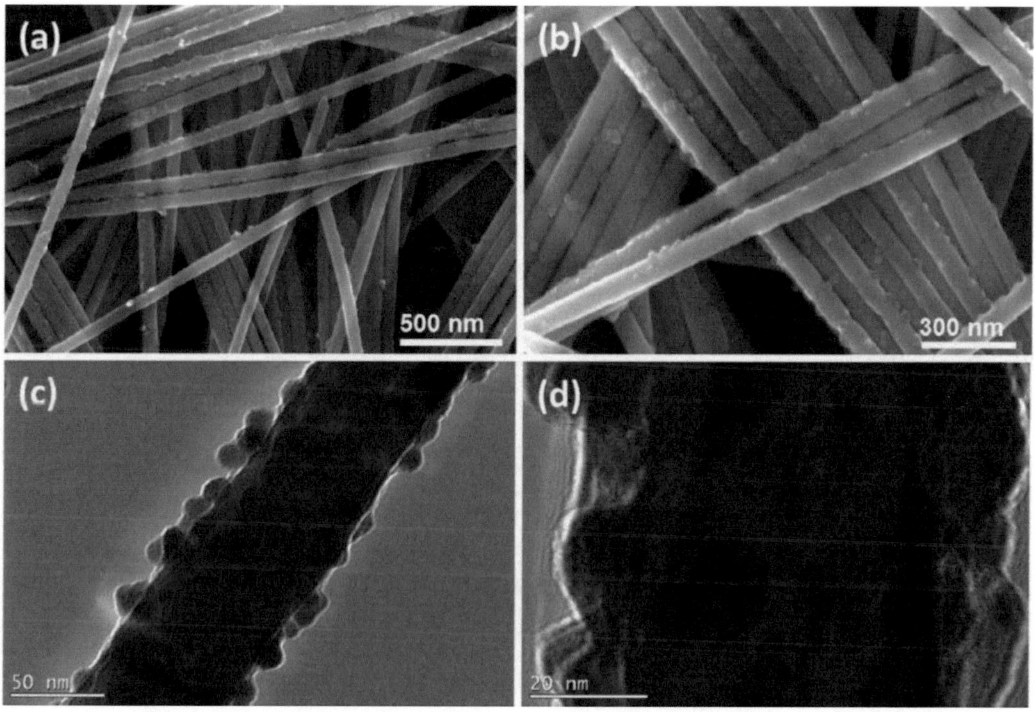

Figure 6. SEM images of nanoparticle-decorated NWs shown in (a) and (b). TEM images of the NWs at different resolutions are presented in (c) and (d), which confirm the successful fabrication of Co nanoparticle-decorated Ni nanowires with diameter of ~60 nm.

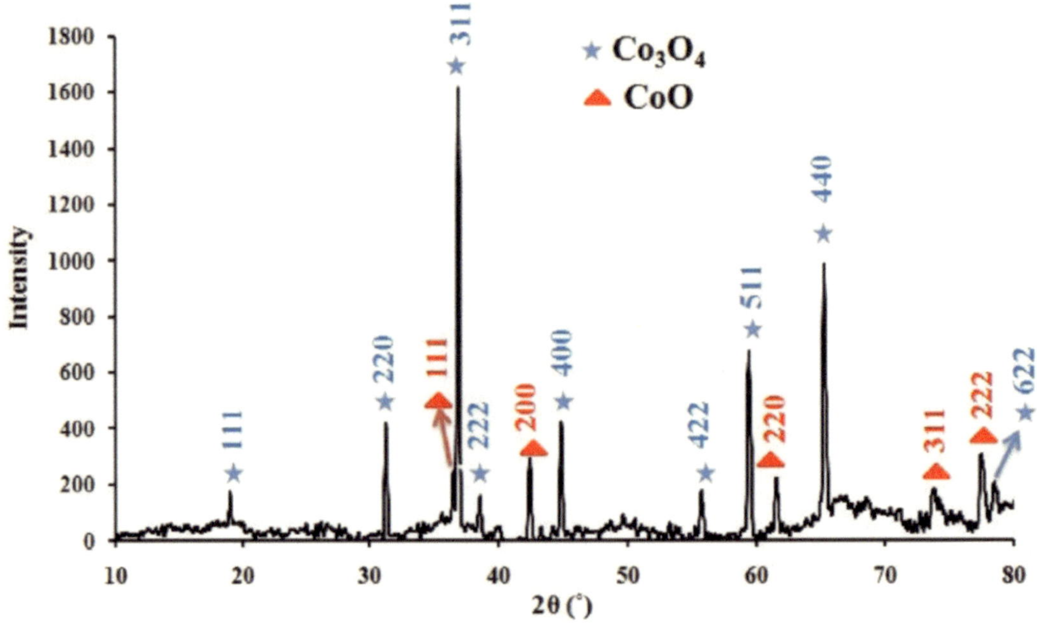

Figure 7. XRD pattern of CoO/Co_3O_4 nanoparticles.

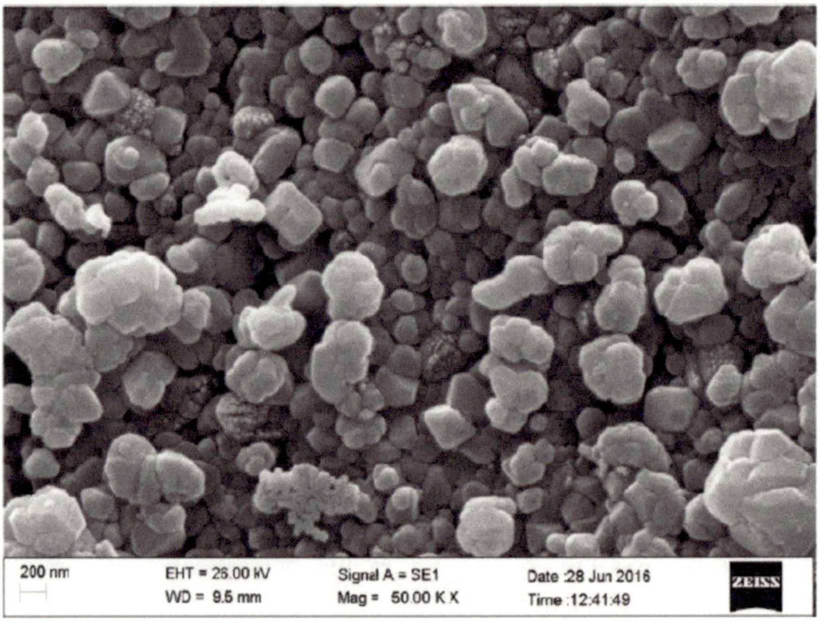

Figure 8. SEM image of CoO/Co_3O_4 nanoparticles.

It was shown, in another work in 2017, a novel dry coating technology for Co/Al_2O_3 catalyst synthesis without solvents and heating treatment. The dry-coated catalysts were presented as comparable to conventional chemical impregnated catalysts. **Figure 9** shows a spherical shape of Co/Al_2O_3 [44].

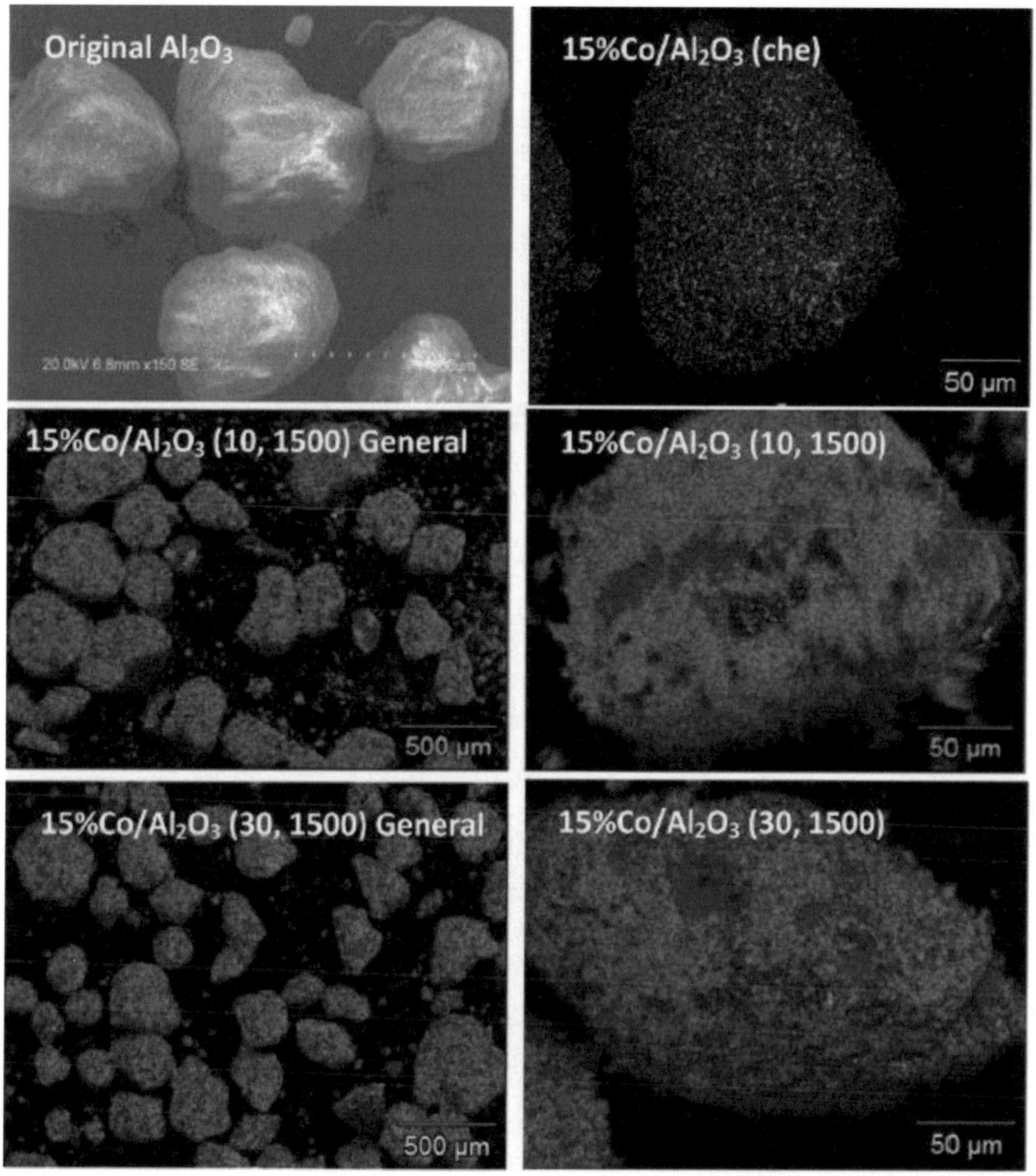

Figure 9. Morphology of the catalysts prepared in "Picomix" and by conventional impregnation: Al (blue color) and Co (green color).

Przyojski et al. synthesized two new complexes of cobalt(II) with 7-azaindole. They have Co(II) in a distorted tetrahedral environment. Asymmetric units of complexes are shown in **Figures 10** and **11** [45].

Four novel complexes $[Co(H_2O)_4(sul)_2]\ 1, [Co(2-ampy)_2(sul)_2]_2, [Co(H_2O)_2(1,10-phen)(sul)_2]_3$ and $[Co(2,9-dimephen)(sul)_2]_4$ $(sul = sulindac, 2-ampy = 2-aminopyridine,\ 1,10-phen = 1,10-phenanthroline$ and $2,9-dimeph = 2,9-dimethyl-1,10-phenanthroline$ were prepared and characterized by Shalash and Abu Ali [16]. This novel complexes were evaluated to determine activity against Gram-positive (*Staphylococcus epidermidis, S. aureus*) and Gram-negative (*Bordetella, Escherichia coli*) bacteria and Yeast species (*Saccharomyces* and *Candida*) using agar well-diffusion method. The results indicated that only complex 4 have been effective against yeast. Also, all of them have more antibacterial activity against Gram-positive bacteria than Gram-negative bacteria [16].

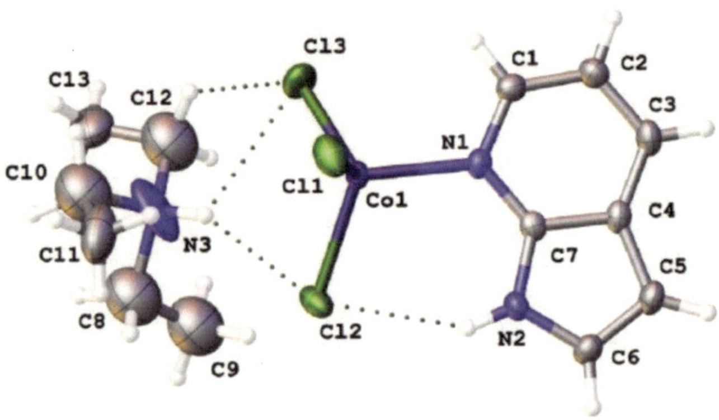

Figure 10. Asymmetric unit of $[(C_2H_5)_3NH]\ [Co(Haza)Cl_3]$, **1**. Thermal ellipsoids are shown at 50% probability. Color code: *C, gray; Cl, green; Co, dark blue; H, white; N, blue.*

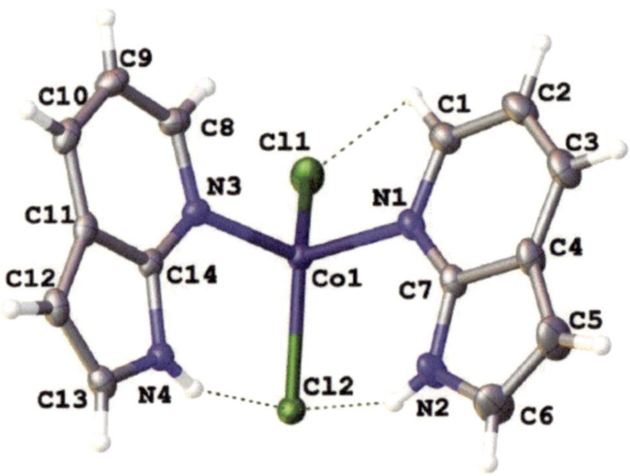

Figure 11. Asymmetric unit of $[Co(Haza)_2Cl_2]$, **2**. Thermal ellipsoids are shown at 50% probability. Color code: *C, gray; Cl, green; Co, dark blue; H, white; N blue.*

Hassanzadeh et al. investigated the cobalt Schiff base complex-modified CPE containing cationic surfactant. This complex could increase resolution and selectivity of voltammetric responses of DA and AA. Also, it was challenging to distinguish the voltammetric peaks of DA and AA. It obtained the better of resolution than previous reported works. This makes it suitable for simultaneous detection of these compounds. Moreover, it was simple preparation, acceptable selectivity and sensitivity, had low-detection limit and reproducibility. All of them made the prepared system very effective in manufacturing [46].

$[(Co(bdmpzm)_2(NCS)_2](bdmpzm = bis(3, 5 - dimethylpyrazol - 1 - yl\ methan$ was synthesized and was integrated with single-walled carbon nanotubes (SWCNT) and nafion on a screen printed carbon electrode (SPCE) by Nagles et al. [47]. They investigated its electrocatalytic activity for dopamine (DA) oxidation. The measurements were made with or without the cobalt complex and compared with the modified electrode. DA in human urine was measured by obtained new sensor in the range of 83, 0–93, 0. Owing to cobalt complex in this electrode, there was an increase in anodic current and change in oxidation potential for DP. Also, unlike other conventional methods, analysis had been done in short time [47].

Studies on cobalt are continuing due to the fact that they have a wide variety of functions and many applications, especially in health sector. Unfortunately, cobalt derivatives have not been studied as pharmaceutical, yet. Up to now, the only cobalt-based drug is Doxovir that is Co(III) Schiff base complex and its mechanism is not also completely understood [36].

5. Toxicity of cobalt

Physical contact with cobalt is not the only way to expose to a substance. You may be exposed by lot of routes such as breathing, eating or drinking the substance, or by skin contact. Metal poisonings occur because of different reasons. For example, cobalt toxicity depends on oral intake and inhalation. Cobalt exposure related to using vitamin B12 is considered as there is low toxicity due to fast rate of renal excretion and limited oral intake. Mostly, absorbed cobalt (50–88%) eliminated by renal excretion rest by feces.

Just as other metals, cobalt is also a multiple organ poison. $CuSO_4$ inhibits many key enzyme systems and prevents the beginning phase of protein synthesis.

It is brought to attention that cobalt may inhibit aerobic metabolism with increasing glycolysis while $CoCl_2$ reduce the consumption of oxygen [48].

Cobalt poisoning causes gastrointestinal distress and heart failure. Chelating agents are used in the treatment of those poisonings [49].

Cobalt also prevents Krebs cycle by generating reduced α-lipoic acid. Besides, cobalt salts inhibit dihydrolipoic acid by forming complex with dihydrolipoic acid sulfhydryl groups.

Result of this reaction, it is not enough pyruvate convert to acetyl co-A and α-ketoglutarate convert to succinyl-coA.

Moreover, $CoCl_2$ inhibit tyrosineiodinase enzyme. This causes the reduction of triiodothyronine (T3) and thyroxine (T4) [48].

6. Conclusion

In this chapter, chemical properties of cobalt and some new cobalt compounds were discussed. It deals with the progress of cobalt chemistry. Cobalt is substantial in both chemical reactions and within many compounds. Some of them are heterocycle reactions, cobalt-based catalyst and cobalamin. Cobalt studies are continuing due to the fact that they have a wide variety of functions and many applications, especially in pharmaceutical technology. These studies caused appearance of medicinal bioorganometallic chemistry and expanding of medical inorganic chemistry. The studies so far in this area carried out have enabled producing unknown and difficult reactions and they will be continued to be responsible for them. This survey of the recent literature illustrates the fact that many different approaches on cobalt and new cobalt compounds are being used in many different areas. For example, one of them is that many different new creative approaches are being taken toward the design of innovative metal-based anticancer drugs.

Author details

Yasemin Yildiz

Address all correspondence to: yyildiz@sakarya.edu.tr

Vocational School of Health Services, Sakarya University, Sakarya, Turkey

References

[1] Davis JR. Nickel, cobalt, and their alloys:2000. Cobalt market review. ASM International. 2000:2009

[2] Kumar DN, Garg BS. Some new cobalt(II) complexes. Journal of Thermal Analysis and Calorimetry. 2002;**69**:607-616

[3] Darton Commodities Ltd. Cobalt Market Report [Internet]. July 2013. [Accessed: Jul 6, 2017]

[4] Tezcan R, Tezcan H. Metaller Kimyası. 1st ed. Ankara: Nobel Yayınları; 2007. p. 288

[5] Gündüz T. Yarı-Mikro Kalitatif Analiz. 2nd ed. Ankara: Ankara Univercity; 1978. p. 243

[6] Baykut F. Anorganik Kimya Praktikumu. 4th ed. İstanbul: İstanbul Univercity; 1974

[7] Zhang S, Chen P, Goo S. Synthesis and Characterization of New M- Trizole Complexes (M=Co,Cu, Zn), Journal of Inorganic and General Chemistry. 2009;**635**(3):537-543

[8] Cody JA, Dalecky LM, Juillerat CA, Alexander GCB, Shylanski LC. Ionothermal synthesis of new cobalt and chromium thiophosphate anions $[Co(Co(P3S8)_2)_2]_4$ and $[Cr(P3S9)_2]_3$. Polyhedron. 2016;**114**:399-402. DOI: 10.1016/j.poly.2016.02.015

[9] Scott WW. Standart Methods of Chemical Analysis I. 5th ed. New York; 1939;(2):2617

[10] Özbek A, Kınayoğlu N. Anorganik Sınai Kimya. 1st ed. İstanbul: Milli Eğitim Basımevi; 1980. p. 282

[11] Dull CE, Metcalfe HC, Williams JE. Modern Chemistry, Holt, Rinehard and Winston, Inc, New York; 1962

[12] Asgharpour Z, Farzaneh F, Abbasi A. Synthesis, characterization and immobilization of a new cobalt(II) complex on modified magnetic nanoparticles as catalyst for epoxidation of alkenes and oxidation of activated alkanes. The Royal Society of Chemistry. 2016;**6**:95729-95739. DOI: 10.1039/c6ra18154f

[13] Wu X-F, Beller M. Economic Synthesis of Heterocycles: Zinc, Iron, Copper, Cobalt, Manganese and Nickel Catalysts. Cambridge: Royal Society of Chemistry; 2014. p. 512

[14] Ingersoll JC, Mania N, Thenmozhiyal JC, Muthaiah A. Catalytic hydrolysis of sodium borohydride by a novel nickel–cobalt–boride catalyst. Journal of Power Sources. 2007;**173**:450-457. DOI: 10.1016/j.jpowsour.2007.04.040

[15] Bruijnincx PCA, Sadler PJ. New trends for metal complexes with anticancer activity. Current Opinion in Chemical Biology. 2008;**12**:197-206. DOI: 10.1016/j.cbpa.2007.11.013

[16] Shalash AM, Abu Ali HI. Synthesis, crystallographic,spectroscopic studies and biological activityof new cobalt(II) complexes with bioactivemixed sulindac and nitrogen-donor ligands. Chemistry Central Journal. 2007;**11**:40. DOI: 10.1186/s13065-017-0268-2

[17] Bauer HF, Drinkard WC. A General Synthesis of Cobalt(III) Complexes: A new intermediate, J. Am. Chem. Soc., Oct 5, 1960;**82**(19):5031-5932

[18] Shaabani A, Farhangi E, Rahmati A, Aerobic oxidation of alkyl arenes and alchols using cobalt (II) phtalocyanine as a catalyst in 1-butyl-3-methy-imidazolium bromide, A. Applied Catalysis, A: General. 2008;**338**:14-19

[19] Sujandi FA, Prasetyanto SCH, Park SE. Styrene Epoxidation over Cobalt Cyclam Immobilized SBA-15 Catalyst, Bulletin of the Korean Chemical Society. 2006;**27**:1381-1385

[20] Lopez-Sandoval H, Londono-Lemos ME, Garza-Velasco R, Poblano-Melendez I, Granada-Macias P, Gracia-Mora I, Barba-Behrens N. Synthesis, structure and biological activities of cobalt(II) and zinc(II) coordination compounds with 2-benzimidazole derivatives. Journal of Inorganic Biochemistry. 2008;**102**:1267-1276

[21] Bhattacharya PK. Metal Ions in Biochemistry. India: Alpha Science International Ltd.; 2005. p. 217 DOI: 1-84265-240-0

[22] Gaëlle DSY, Yufanyi DM, Jagan R, Agwara MO. Synthesis, characterization and antimicrobial properties of cobalt(II) and cobalt(III) complexes derived from 1,10-phenanthroline with nitrate and azide co-ligands. Cogent Chemistry. 2016;**2**:1-16. DOI: 10.1080/23312009.2016.1253201

[23] Mafatle T, Nyokong T. Use of cobalt(II) phthalocyanine to improve the sensitivity and stability of glassy carbon electrodes for the detection of cresols, chlorophenols and phenol. Analytica Chimica Acta. 1997;**354**:307-314

[24] Das BK, Clark JH. A novel immobilised cobalt(III) oxidation catalyst. The Royal Society of Chemistry. 2000;(7):605-606. DOI: 10.1039/b000535p

[25] Park J-I, Cheon J. Synthesis of "solid solution" and "core-shell" type cobalt-platinum magnetic nanoparticles via transmetalation reactions. American Chemical Society. 2001;**123**:5743-5746. DOI: 10.1021/ja0156340

[26] Nikitenko SI, Koltypin Y, Palchik O, Felner I, Xu XN, Gedanken A. Angewandte Chemie. 2001;**40**(3):447-449

[27] Kobayashi Y, Horie M, Konno M, Rodrıguez-Gonzalez B, Liz-Marza LM. Preparation and properties of silica-coated cobalt nanoparticles. The Journal of Physical Chemistry. B. 2003;**107**:7420-7425. DOI: 10.1021/jp027759c

[28] Li T, Yang S, Huang L, Benxi G, Youwei D. A novel process from cobalt nanowire to Co_3O_4 nanotube. Nanotechnology. 2004;**15**:1479-1482 DOI: 0957-4484/04/111479+04

[29] Podunavac-Kuzmanović SO, Cvetković DM, Gordana SĆ. Antimicrobial activity of cobalt(II) complexes with 2-Aminobenzimidazole derivatives. Apteff. 2004;**35**:1-280

[30] Song Y, Modrow H, Henry LL, Saw CK, Doomes EE. Microfluidic synthesis of cobalt nanoparticles. Chemistry of Materials. 2006;**18**:2817-2827. DOI: 10.1021/cm052811d

[31] Salavati-Niasari M, Davar F, Mazaheri M, Shaterian M. Preparation of cobalt nanoparticles from [bis(salicylidene)cobalt(II)]–oleylamine complex by thermal decomposition. Journal of Magnetism and Magnetic Materials. 2008;**320**:575-578. DOI: 10.1016/j.jmmm.2007.07.020

[32] Hamil AM, Khalifa KM, AL-Houni A, El-ajaily MM. Synthesis, spectroscopic investigation and antiactivity activity of Schiff Base complexes of cobalt(II) and copper(II) ions. Rasāyan. Journal of Chemistry. 2009;**2**:261-266

[33] Kumar U, Chandra S. Biological active cobalt(II) and nickel(II) complexes of 12-membered Hexaaza [N6] macrocyclic ligand synthetic and spectroscopic aspects. E-Journal of Chemistry. 2010;**7**(4):1238-1245

[34] Pannu APS, Kapoor P, Hundal G, Kapoor R, Corbella M, Aliaga-Alcalde N, Hundal MS. Magneto-structural studies of two new cobalt(II)-N,N-diisobutylisonicotinamide compounds: $[CoLCl_2]_n$ and $[Co(L)_2(H_2O)_4][CoLBr_3]_2{\cdot}2H_2O$. Dalton Transactions. 2011;**40**:12560-12569. DOI: 10.1039/C1DT10991J

[35] Ke G, Yoshikai N. Low-Valent cobalt catalysis: New opportunities for C–H functionalization. Accounts of Chemical Research. 2014;**47**:1208-1219. DOI: 10.1021/ar400270x

[36] Morcelli SR, Bull ÉS, Terra WS, Moreira RO, Borges FV, Kanashiro MM, Bortoluzzi AJ, Maciel LLF, Almeida JC d A, Horn Júnior A, Fernandes C. Synthesis, characterization

and antitumoral activity of new cobalt(II)complexes: Effect of the ligand isomerism on the biological activity of the complexes. Journal of Inorganic Biochemistry. 2016;**161**:73-82. DOI: 10.1016/j.jinorgbio.2016.05.003

[37] Dong S, Shi L, Zhang S-L, Zhao X-H, Wu D-Q, Wei X-Q, Wang X-Y. Syntheses, structures, and magnetic properties of three new chain compounds based on a pentagonal bipyramidal Co(II) building block. CrystEngComm. 2016;**18**:4150-4157. DOI: 10.1039/c5ce02594j

[38] Azizolla Beheshti FS, Hashemi F, Motamedi H, Mayer P, Bruno G, Rudbari HA. Synthesis, structural characterization, antibacterial activity and computational studies of new cobalt(II) complexes with 1,1,3,3-tetrakis (3,5-dimethyl-1-pyrazolyl)propane ligand. Journal of Molecular Structure. 2016;**1123**:225-237. DOI: 10.1016/j.molstruc.2016.06.037

[39] Adam R, Bheeter CB, Cabrero-Antonino JR, Junge K, Jackstell R, Beller M. Selective hydrogenation of nitriles to primary amines by using a cobalt phosphine catalyst. ChemSusChem. 2017;**10**:842-846. DOI: 10.1002/cssc.201601843

[40] Ko YN, Choi SH, Kang YC. Hollow cobalt selenide microspheres: Synthesis and application as anode materials for Na-ion batteries. ACS Applied Materials & Interfaces. 2016;**8**:6449–6456. DOI: 10.1021/acsami.5b11963

[41] Mondal R, Sahoo S, Rout CS. Mixed nickel cobalt manganese oxide nanorods for supercapacitor application. American Journal of Engineering and Applied Sciences. 2016;**9**(3):540-546. DOI: 10.3844/ajeassp.2016.540.546

[42] Maaz K, Duan JL, Karim S, Chen YH, Yao HJ, Mo D, Sun YM, Liu J. Magnetic properties of nickel nanowires decorated with cobalt nanoparticles fabricated by two step electrochemical deposition technique. Materials Chemistry and Physics. 2016;**182**:466-471

[43] Montazerozohori M, Masoudiasl A, Farokhiyani S, Joohari S, McArdle P. Sonochemical synthesis of a new cobalt(II) complex: Crystal structure, thermal behavior, Hirshfeld surface analysis and its usage as precursor for preparation of CoO/Co_3O_4 nanoparticles. Ultrasonics Sonochemistry. 2017;**38**:134-144

[44] Mengnan L, Fatah N, Khodakov AY. New shearing mechanical coating technology for synthesis of alumina-supported cobalt Fischer–Tropsch solid catalysts. Journal of Materials Chemistry A. 2017;**5**:9148-9155

[45] Przyojski JA, Arman HD, Myers NN, Walmsley JA. Synthesis and characterization of two new mononuclear complexes of cobalt(II) with 7-Azaindole. Journal of Chemical Crystallography. 2017;**47**:22-29

[46] Hassanzadeh N, Zare-Mehrjardi HR. Selective electrochemical sensing of dopamine and ascorbic acid using carbon paste electrode modified with cobalt Schiff Base complex and a surfactant. International Journal of Electrochemical Science. 2017;**12**:3950-3964. DOI: 10.20964/2017.05.07

[47] Nagles E, Ibarra L, Llanos JP, Hurtado J, Nagles OG-BE, Ibarra L, Llanos JP, Hurtado J, Garcia-Beltrán O. Development of a novel electrochemical sensor based on cobalt(II) complex useful in the detection of dopamine in presence of ascorbic acid and uric acid. Journal of Electroanalytical Chemistry. 2017;**788**:38-43. DOI: 10.1016/j.jelechem.2017.01.057

[48] Hoffman RS, Nelson LS, Howland MA, Lewin NA, Flomenbaum NE, Goldfrank LR. Goldfrank's Manual of Toxicologic Emergencies. New York: Mc-Graw Hill; 2007

[49] Bertini I, Gray HB, Lippard S, Valentine J. Bioinorganic Chemistry. Mill Valley, California: University Science Books; 1994

Chapter 2

Synthesis, Processing, and Characterization of the Cobalt Alloys with Silicon Addition

Petrica Vizureanu, Mirabela Georgiana Minciuna,
Gianina Iovan and Simona Stoleriu

Additional information is available at the end of the chapter

http://dx.doi.org/10.5772/intechopen.70886

Abstract

The composition of the cobalt alloys contains only noncytotoxic elements (Cr, Si, and Mo) that ensure its biocompatibility, and consequently, the development and proliferation of cells at the implant/tissue interface. The cobalt alloy has an original composition with silicon addition and the proportion of the alloying elements was established so as to ensure a high biocompatibility and adequate physical-chemical characteristics for it to be used in various applications. Silicon is known to be a metal with a high biocompatibility; it can replace noble/non-noble metals in commercial alloys, thereby excluding the occurrence of any toxic corrosion products. We chose it as an alloying element because it confers good casting properties, has double role as hardener and oxidant, ensures an increase in the resistance to tear, and offers a proper fluidity in the liquid phase.

Keywords: cobalt, elaboration, microstructural characterization, tensile tests, electrochemical impedance spectroscopy

1. Introduction

This chapter aims to provide in-depth study and analysis of various synthesis methods, processing techniques, and characterization of cobalt that will lead to its increased application in the technology. The information provided is addressed to engineering professionals, medicine, manufacturers, and material scientists.

Our research is a cobalt-based alloy with an original composition that includes nontoxic and nonallergic Si as an alloy component, which endows the alloy with a high chemical resistance and a high biocompatibility. Our cobalt-based alloy is intended for medical applications.

Metallic dental alloys usually have very good durability, mechanical resistance, and biocompatibility characteristics. Dental alloys became more diverse in time, depending on the technology applied and the characteristics required for a specific type of dental prosthesis.

Cobalt-based alloys are frequently used to make the metallic framework of prostheses or to refurbish deteriorated ones due to their biocompatibility, their outstanding mechanical resistance, and their higher elastic modulus compared to other conventional alloys. The elastic modulus of titanium is inferior compared to that of cobalt-based alloys, its mechanical resistance is low, and the size of the prostheses made from it is larger.

The global concern to improve the classical technologies of execution both of the implants, and also the biomaterials from which are made, aims in the promotion of a new multifunctional implants, with best performances for a long time. About the materials for implants, the actual tendencies both in medical practice and also in research follow the utilization of materials with advanced biological and biomechanical characteristics, with advantages both in terms of biocompatibility with human tissue, and also the avoidance the risks of the infections or the rejection after implanting [1].

The classical alloys used such as biomaterials, the stainless steels, and titanium base alloys present disadvantages like elasticity modulus greater than biological materials (bone, conjunctive tissue), small corrosion resistance, or low biocompatibility with human tissues.

The most recent researches in medical domain show that titanium is a material with high biocompatibility, but his alloying with various metals to improve the mechanical characteristics is not always beneficial. It is the case of vanadium and nickel, which are classified as toxic elements, with carcinogen effects and aluminum, which are present in most of titanium commercial alloys, and which has shown a causal relationship with neurotoxicity [2].

The new type of implant is the zirconium and it presents numerous advantages compared to titanium; these affirmations have been sustained by dental medics. The zirconium implants are integrated better in soft tissues (in gingival), present high resistance, look like natural teeth, do not produce allergies, do not lead electrical current and heat, and are more easily accepted by human body, but the bigger disadvantage is about the price, his utilization is thus limited.

Cobalt-base alloys, like Vitalium, Vicalloy, "C" alloy, Wironit Extrahart, became the most used in manufacturing medical implants. The presence of chrome, silicon, and molybdenum has made the cobalt-base alloy to be considered the most healthy, safety, nontoxic, and nonallergic alloys, the fact demonstrated by laboratory analyses. These alloys present good cast properties, corrosion resistance, nonallergic properties, compatibility with human tissue, and high viability of cell, des-oxidation role.

Actually, researches have been developed in United States of America, European Union, Japan and China about the influence of the presence of cobalt alloying elements on the fibroblasts and osteoclasts cells from the appropriate tissue soft implants, more precisely at the interface implant/tissue [3]. The obtained results highlighted that alloying elements as nickel, aluminum, vanadium, and titanium are toxic, having a carcinogen character and a causal relationship with neurotoxicity and senile dementia, like Alzheimer.

2. Characterization of the cobalt alloys with silicon addition

The objective of the study was the elaboration of new performing alloys based on cobalt that consists in the identification of new composition, increasing the silicon content (99.99% purity) up to 10%.

Remelting was performed in electric arc furnace with an argon atmosphere that had been previously vacuumed at 10^{-4} mbar [4]. Under the influence of the advanced vacuum in the chamber and in the crystallizer, a strong degassing occurs leading to a low gas content (nitrogen and hydrogen) in the final remelted alloy.

The furnace was used to remelt the alloy with its electric arc between the Φ 6.5 mm thoriated wolfram electrode and the metallic load. The crucible copper of the water-cooled electric arc remelting installation has cavities of various shapes (**Figure 1**) where the raw ingots were cast. To obtain a superior chemical and structural homogeneity, we remelted the alloy for seven times.

The technical characteristics of the installation are as follows [6]:

- water cooled 304 L stainless steel vacuum chamber with double walls;
- the crucible of the working chamber made of aluminum;
- loading is easy due to the articulated back of the bell;

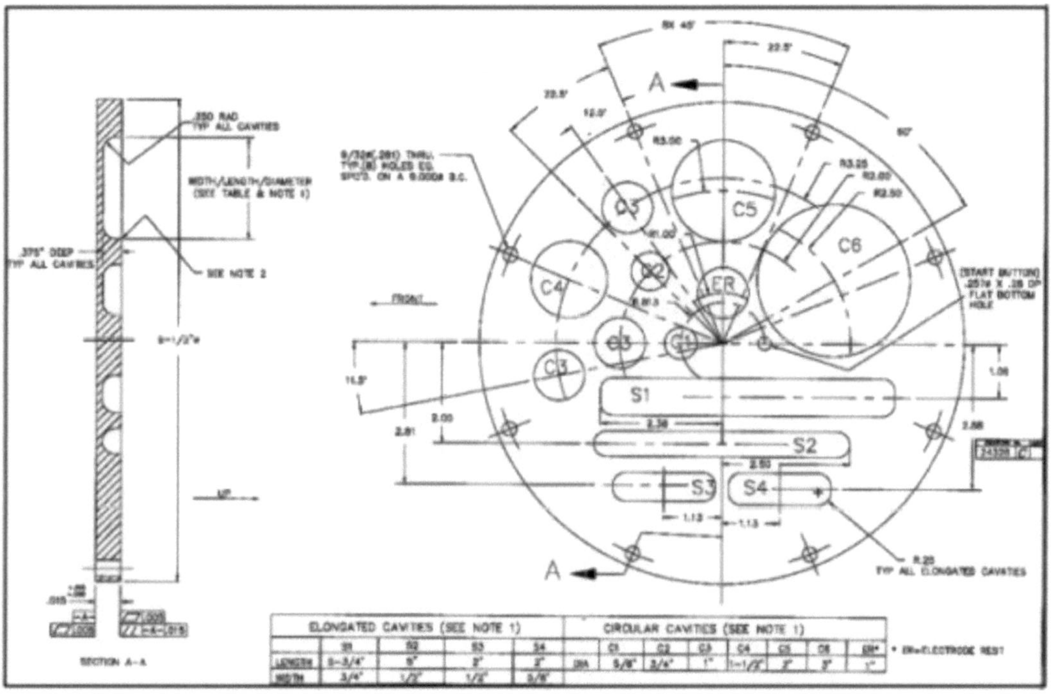

Figure 1. Crucible design of VAR furnace [5].

- visualization: the crucible and the electrode may be observed through the 4″ (101.6 mm) window on the face of the furnace; behind there is a separate 1″ (25.4 mm) window for lighting;
- the visiting window is fit with an adjustable welding view-port to protect the eyes during melting;
- the copper crucible has a series of cavities and is fixed to the water-cooled base of the furnace;
- power supply cables to the electrode and the base of the furnace are water cooled;
- the bell-shaped working chamber is mounted on a frame that encircles the welding device, to reduce the surface of the furnace;
- a thoriated wolfram electrode (6.35 mm DIA. × 76.2 mm) is mounted centrally and may be moved radially so as to cover the entire surface of the crucible;
- a protective sleeve is covering the electrode to prevent the electrocution hazard during functioning;
- the advanced vacuuming system (0.013–0.0013 N/m^2) consists of a preliminary vacuuming pump and a diffusion pump;
- oxygen analyzer with a measuring domain of 0–10,000 ppm;
- additional loading mechanism during melting;
- release gauge at 13,780.2 N/m^2, fit to the working chamber to maintain a positive pressure inside, preventing gasses to enter the furnace.

For this type of furnace, the crucible in which the alloy is melted and produced is of high importance. The crucible of the furnace has oblong and circular cavities in which we obtained casting samples in standard test-piece sizes, necessary to characterize the properties of the alloys under research.

The total quantity of gas that can be exhausted during vacuum melting largely depends on the chosen raw materials, because at low pressures and even at an increase of the carbon content, the stable oxides may only partially get decomposed. Therefore, it is required that the materials intended to be melted contain minimal quantities of elements with a high affinity to oxygen. The metallic load from which to obtain biocompatible alloys of the CoCrMo system must be of very good quality, with a low content of phosphorus and sulfur, they must be degreased and properly prepared mechanically. During the first phase of the melting process, the maximum quantity of gas is released, 3–4 times more than that released during the refinement phase of the melting process. The degree of purity of the elements that compose the alloy is of over 99%.

The elements dosed in equimolecular proportions are introduced in an order dictated by the type of the cobalt-based alloy to be made, in the cavity on the crucible copper, from which one can subsequently make test pieces for various kinds of investigations.

Our experimental studies were performed on standard test pieces and focused on surface, chemical, and physical-structural characterization (cavity C4), mechanical and clinical study on the biocompatibility of the alloy to be patented (cavity S2).

2.1. Technological workflow

The technological workflow to produce the CoCrSiMo alloy involved the use of a remelting furnace presented in **Figure 1** and it comprised the following operations:

- preparation of the raw materials (a CoCrMo pre-alloy and Si) by cutting to sizes adequate to the crucible and cleaning them in an ultrasound bath followed by degreasing with volatile organic solvents;
- raw material dosage by weighing, according to batch calculations;
- loading the raw materials into the crucible of the furnace;
- vacuuming the chamber, obtaining an inert controlled atmosphere (Ar) and pressure level inside the melting chamber;
- melting;
- gravitational casting;
- cooling and extraction of the solidified ingot from the crucible.

2.1.1. Raw materials

Designing a technology to obtain a cobalt-based alloy to be used in medical applications was based on our target to improve the quality of existing alloys, by using advanced analysis techniques that would confirm an improvement in certain properties. The vacuum arc remelting (VAR) furnace ensures the lack of contaminants in the alloy, and consequently, its quality is directly influenced by the purity of the raw materials used.

Considering the special destination of the alloy, it is absolutely necessary that raw materials with minimum impurities can be used, ones that would comply with the following quality requirements:

- prefabricated CoCrMo alloy with the following composition:

Co	Cr	Mo	C	Si	Mn
65%	29%	5%	0.40%	0.35%	0.25%

- 99.99% pure silicon.

2.1.2. Preparation of the raw materials

The preparation of the materials used to make our alloy consisted in cutting to pieces by electro-erosion. Electro-erosion is based on the erosive effects of electrical arcs in impulse, repeatedly applied between the material to be processed and an electrode, to remove the surplus material, and it allows cutting the metallic material into precise pieces (precise tolerance).

After processing by electro-erosion, the materials were chemically degreased, both to avoid contamination of the protective atmosphere inside the furnace and to avoid impurities in the resulted alloy.

2.1.3. The composition of the metallic load

To calculate the metallic load (dosage), one must take into account the theoretical degrees of assimilation of the elements in the melt mix and the possible losses by vaporization during the metallurgic process in vacuum or in argon-controlled atmosphere [4].

The losses are estimated based on data from relevant publications, on the degree of oxidation of the materials in the load, on the characteristic of those elements, on their position in the series of electrochemical potentials, on the characteristics of the melting device, and on the information from researches on the elaboration process.

The metallic load from which to obtain biocompatible alloys of the CoCrMo system must be of very good quality, with a low content of phosphorus and sulfur, they must be degreased and properly prepared (granulometrically).

The degree of purity of the elements that compose the alloy is of over 99%. The elements dosed in equimolecular proportions are introduced in an order dictated by the type of the cobalt-based alloy to be made, in the cavity on the water-cooled crucible copper.

The size of the batches was chosen so as to allow each sample to be processed after elaboration for various tests in regard to surface, physical, chemical, and structural characterization and for the clinical test of biocompatibility (cavity C4, **Figure 1**) and for mechanical properties (cavity S2, **Figure 1**).

2.1.4. The crucible loading

The losses by vaporization were low, because the raw materials we used did not contain impurities and elaboration was performed in a controlled atmosphere, in vacuum, and argon. The dosage of raw materials was done in progressive order according to specific weight; elaboration time was relatively short, which limited the evaporation losses to a minimum.

2.1.5. Achieving pressure levels

In order to make highly pure biocompatible alloys from the CoCrMo class, the working chamber was properly prepared by successive vacuuming operations and by argon purging for three times. Vacuuming was done by the preliminary vacuuming installations and the diffusion pumps that may ensure pressure levels of app. $3.5–4 \times 10^{-4}$ mbar. For purging, we used 99.99% pure argon 5.3. Those operations ensure maximum 40–60 ppp oxygen content in the working chamber. The last stage in the process was argon purging the working chamber and establishing working pressure levels slightly above atmospheric pressure.

2.1.6. The melting of the metallic load

The process of making alloys from the CoCrMo class consists in melting materials under the action of an electric arc created between the thoriated wolfram electrode and the metallic load. The electric arc is gradually conducted over the entire surface of the load, aiming an in-depth melting thereof. Adjustments in the power of the arc are done by manual controls.

The solidification of the obtained alloy occurs very fast, due to the intense water cooling of the crucible. Next, we remelted the samples seven times and returned them into the cavity from the crucible copper, so as to ensure a state of complete alloying and to improve the degree of chemical and thermal homogenization of the ingots. The entire elaboration process was performed by electric arc remelting in argon-controlled atmosphere.

The melting conditions in the installation were adapted to comply with the type of alloy to be obtained.

The thoriated wolfram electrode was moved during melting at a distance of approximately 6.35 mm from the electrode on the crucible copper and the electric arc covered the entire surface of the load so as to ensure a complete homogeneity.

After finishing melting and solidification, we obtained ingots with masses almost constantly matching the mass of the load we introduced into the VAR furnace.

2.1.7. Cooling the metallic test pieces

The alloy cooled almost instantly under the influence of the forced water cooling of the crucible copper (the crystallizer of the installation). The samples we obtained (ingots) had various shapes according to the shape of the cavity in the crucible copper. Those samples (test pieces) were analyzed to determine their chemical composition, their physical, chemical, and mechanical properties.

2.1.8. Extracting the metallic test pieces

After solidification, which occurred in the cavity from the crucible copper, the installation was kept for approximately 10 min in the argon-controlled atmosphere to avoid their oxidation during removal at high temperatures.

Afterwards, we removed the argon source from the installation and returned it to a regular atmo sphere. The obtained samples were then weighted and marked to determine their properties.

2.1.9. Material balance

The samples obtained were classified according to their weight and during the remelting process, there were no weight losses.

The assimilation efficiency of the chemical alloying components depended on the following factors: the chemical efficiency of the elements in regard to oxygen, the specific mass of the alloying material, the degree of purity, chemical composition, etc.

The metallic load of the obtained cobalt-based alloy was calculated by the analytic method. The calculation of the load for the original alloy took into account, both the possible losses by vaporization and the theoretical degrees of assimilation of the elements in the mix.

2.2. Determination of chemical composition by optical emission spectrometry and EDX

The determination of chemical, structural, physicomechanical, and biocompatibility characteristics has been highlighted by experimental laboratory determinations performed on standard specimens specific to the tests. The methods of analysis that were used (optical and electron microscopy, structural analysis by X-ray diffraction, elasticity test, fractal analysis, hardness tests, corrosion resistance, and biocompatibility tests) are in compliance with the relevant technical standards and metallic alloy norms with applications in medicine [7].

The analyses are comparative with the same types of tests being carried out for both CoCrMo commercial alloy and CoCr alloy variants with different silicon additions.

The analysis were focused to characterize the CSi_k alloys ($k = 4, 5, 6$, and 7), which certify that we choose the appropriate technology, the efficiency of the arc melting furnace being very high (99.97%). The original alloy had a composition very close to the one we initially calculated, the losses registered being insignificant (**Figure 2**).

Using advanced laboratory equipment, analyses of new alloy variants give the assurance of a complete and highly scientific characterization, both from a chemical, structural, physicomechanical, and biological point of view. Major advances and spectacular advances in medical sciences have been due to ongoing efforts by research lab specialists in the field of biomaterials engineering to design and develop new alloys that can best substitute components of the human body, by applying modern reconstructive and repair techniques. The chemical composition, both for the CoCrMo commercial alloy and the developed variants, was achieved by optical emission spectrometry. The determined mass concentrations of elements of cobalt alloys are specified in **Table 1**.

By increasing the percentage of silicon, the alloying elements exhibited lower values, a significant change being to the main element, cobalt. If initially it was 62.10% in the CoCrMo commercial alloy, it reached 56.40% in CSi_7 alloy.

2.3. Microstructural characterization of cobalt alloys

Macroscopic analysis is the first step of a surface structure analysis. This analysis requires a minimum of training and gives information on the nature of the alloy in the CoCrMo

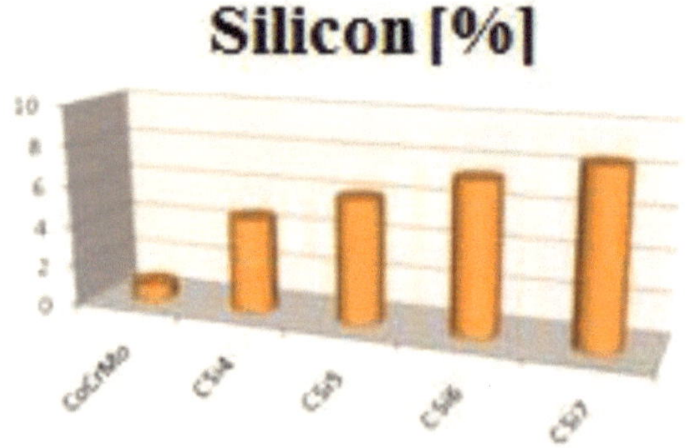

Figure 2. Variations determined by optical emission spectrometry for alloys in the CoCrMo system.

Alloy element	CoCrMo	CSi_4	CSi_5	CSi_6	CSi_7
	(% weight)				
Co	62.10	59.40	58.50	56.70	56.40
Cr	26.79	26.46	26.16	26.23	25.48
Mo	6.00	5.39	5.24	5.29	5.20
Ni	2.90	2.72	2.62	2.84	2.87
Si	0.78	4.64	6.06	7.40	8.45
Mn	0.42	0.38	0.43	0.39	0.38
Fe	0.33	0.43	0.31	0.58	0.43
Others	0.68	0.58	0.68	0.57	0.79

Table 1. Chemical composition of alloys in the CoCrMo system.

system, the peculiarities of the casting structure, the character, and quality of subsequent machining, which confers the shape and final properties, the nature of the break, and its causes [8]. At the same time, the microscopic analysis allows the selection of the areas in the studied specimen, which must then be subjected to a detailed microscopic analysis. Microstructure of alloys used in medical applications is defined by crystallographic orientation, texture, morphology, distribution, size, and number of phases. A Vega Tescan LSH II scanning electron microscope was used to highlight the CoCrMo and CSi_k alloys (k = 4, 5, 6, and 7) (**Figure 3**).

After completion of the elaboration process, the specimens required for structural analysis were processed by electro-erosion, and as methods of preparation we used polishing and revealing of the structure by chemical attack. The specimens of cobalt alloys subjected to microscopic analysis had a cubic shape with the side of 10 mm. The microstructure of metallic materials is a fine construction of the structure, which was highlighted using a metallographic chemical attack with the following composition: 5 ml of HNO_3, 200 ml of HCl, and 65 g of $FeCl_3$.

The SEM microstructures analyzed for commercial alloy and silicon alloy variants of 4, 5, 6, and 7% at the 500× BSE magnification power, show a dendritic structure specific to cobalt alloys cast with the existence of α and β phases, in varying proportions modifiable with increasing the percentage of silicon during casting processes.

From the point of view of the microstructural features, the alloyed with silicon alloy variants exhibit a polymorphic transformation (h.c.—c.f.c.) due to heating/cooling of the cobalt. The experimental research by electronic microscopy confirms the improvement of the properties of cobalt alloys by obtaining a uniform structure with fine eutectic dendritic separations.

2.4. Investigations by X-ray diffraction

X-ray diffraction is the process by which radiation, without wavelength to change, is transformed by interference with the crystal lattice into a large number of observable "reflections" with characteristic spatial directions (**Figure 4**).

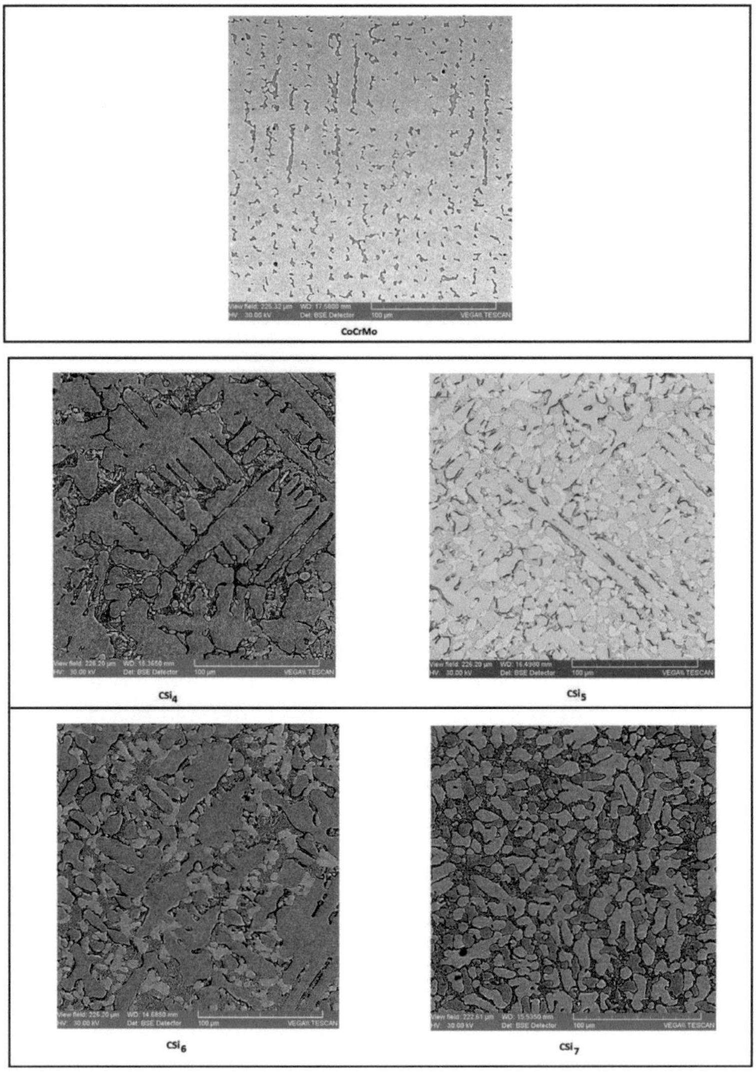

Figure 3. Microstructure of the alloys investigated at a 1000× back scattered electrons (BSE) magnification power.

Indexing of the diffractogram reveals the following: nanocrystalline character and phase composition (major phase, minor phase) of the analyzed sample. Determination of the structural constituents is done by diffractograms in which the dependence between the diffraction radiation intensity and the double diffraction angle is represented.

The determination of the compositional phases was performed by qualitative analysis by X-ray diffraction, performed on the PanalyticalX'Pert PRO MPD X-ray diffractometer. The analysis range 2θ was between 20 and 1000, the step size being 0.0010, and the step time being 3 s/step. One channel proportional detector was used and the analysis being done in the Gonio mode. Cobalt alloy test specimens used for X-ray diffractometry have a 10 mm rectangular section and a 15 mm sample length. Indexing the diffractogram is the association between a peak-peak diffraction and a plane. Diffractometric X-ray analysis confirms the structural changes identified by microscopic analyses and made it possible to accurately determine the phases and structural constituents found in both the commercial alloy (Co_xCr_y, Cr_yMo_z, Cr_xMo_zSi, $Co_xCr_yMo_z$) as well as the elaborated and cast CSi_k alloys (k = 4, 5, 6, and 7).

2.5. Hardness determination of cobalt alloys

The hardness measurements were performed on a universal Wilson Wolpert machine, model 751 N, using a load force of 9807 N and a measurement time of 12 s. The Vickers method was chosen because this is a general method for determining the hardness of metallic materials

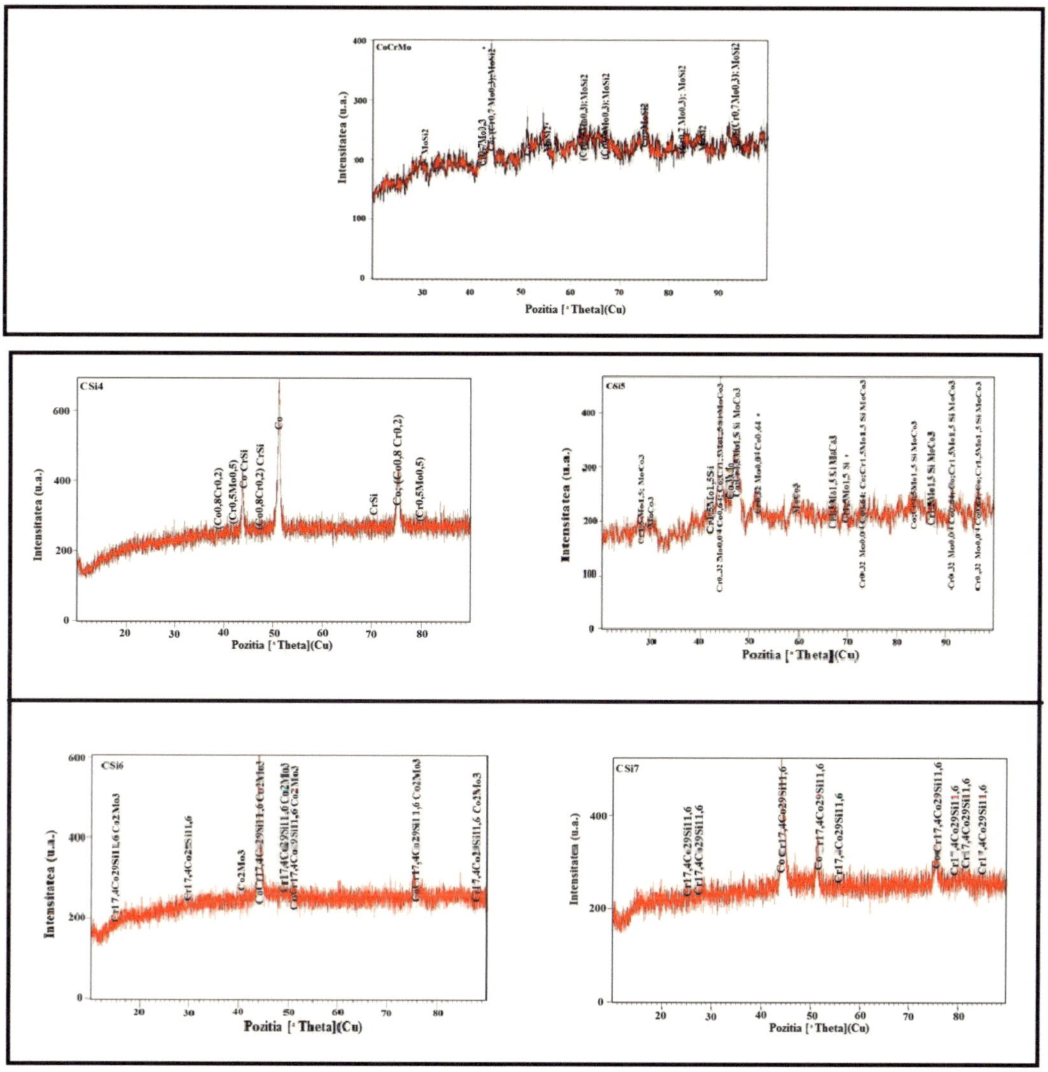

Figure 4. X-ray diffraction spectrums.

and can be used without reservation in the case of cobalt-based experimental alloys [9]. For the accuracy of the results, three determinations were made for each alloy with certain measurement conditions. The hardness determination was performed on samples with two parallel planes, one of the surfaces being prepared by grinding on abrasive paper (**Table 2**).

The hardness measurements made on CoCrMo alloys provide information on mechanical strength, requiring or not the utility of thermal treatments. The silicon input added to the commercial alloy version of the CoCrMo system improved mechanical characteristics, especially hardness, by forming solid cobalt solutions and Cr_ySi and Mo_zSi chemical compounds, also favoring a fine grain structure. In CSi_6 and CSi_7 versions, the hardness increased (639.6 HV), sometimes making mechanical machining impossible, but in terms of corrosion resistance, they could have a major advantage. In this context, a possible processing with the help of a special instrument (extruder cutters) is envisaged.

2.6. Tensile tests

The tensile test of the standard specimens, made of cobalt alloys, was performed on the computer-assisted Instron 3382 test machine. The results obtained of the tensile tests were: elongation, modulus of elasticity, traction resistance, and elongation at break and they provide complete information on the mechanical properties analyzed. In order to be sure of the experimental results obtained, the research was carried out on specimens with specific dimensions to ISO 6892-1: 2009 (E) for both the tensile tests and the machine operating rules [10].

As a following of the tensile test, results have been obtained with respect to elongation, elastic modulus, traction resistance, and elongation at break. From the analysis of the mechanical characteristics variation, depending on the increase of the silicon intake, one can notice an obvious improvement of the properties of the alloys.

The specific elongation of cobalt alloys has close values (≈11%) for the CoCrMo commercial alloy and the CSi_4 and CSi_5 variants. This has also been confirmed by measured hardness values (up to 500 HV, alloys specific for medical applications). By alloying with 6 and 7% Si, CoxMoy and CrzCoxSi combinations (identified by qualitative X-ray diffraction analysis) that radically altered the mechanical properties of the alloys appeared (**Figure 5**).

According to experimental determinations, elongation decreases to below 1%. This is in direct correlation with both microstructural changes (interdendritic agglomerations) and mechanical

Alloy	**CoCrMo**	**CSi_4**	**CSi_5**	**CSi_6**	**CSi_7**
Method	**HV**				
Measured punctual values	423	446	513	633	720
	446	458	484	633	746
	446	471	498	653	772
Average value	438.30	458.30	498.30	639.60	746.00

Table 2. Hardness of cobalt alloys.

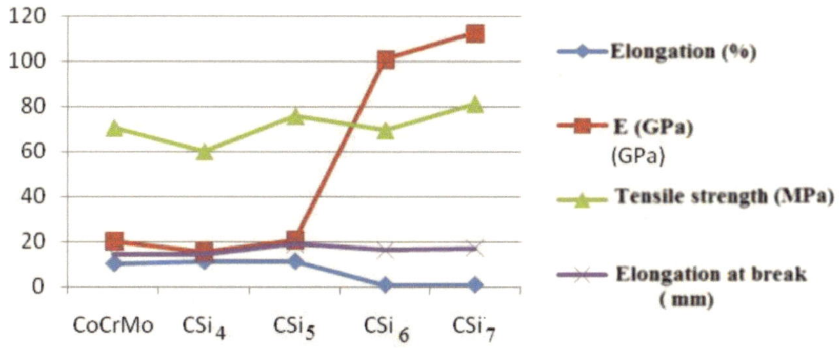

Figure 5. Variation of mechanical characteristics.

strength, due to the excessive increase in hardness to values above 500 HV. The modulus of elasticity determined by tensile tests for the CoCrMo commercial alloy has values close to the experimental alloys CSi_4 and CSi_5 (the modulus of elasticity has values near to steels). The determined modulus of the elasticity modulus guarantees an improvement in the properties of the cobalt alloys providing good deformability. By alloying with silicon of 6 and 7%, the modulus of elasticity increases excessively, this gives the alloy some resistance to the plastic deformation processes.

On the basis of the measurement of the elongation test, it was found that the alloyed silicon alloys exhibited alternating values around the average value of the CoCrMo commercial alloy. From the macroscopic analysis of specimens subjected to tensile tests, both the control sample and variants in the CoCrMo system revealed a fragile fracture of the standard specimens.

2.7. Fractographic analysis

Cobalt alloy test specimens subjected to tensile testing were further investigated by fractal analysis. This analysis offers a complete characterization of mechanical properties, on how to behave in different mechanical actions (elasticity, fragility, breaking strength, creep, etc.) [11]. The examination by electronic microscopy (SEM) was performed according to: SR EN 1321: 2001. For the determination of the fractal analysis, an electronic microscope, Quanta Inspect S, FEI was used. The microscopic analysis was performed using the back scattered electrons (BSE) detector (2D image of the surface, better contrast of the different phases), the 1000× magnification order.

CoCrMo alloys subjected to fractal analysis were cleaned with propanol in an ultrasonic and warm-air bath (not chemically attacked). Measuring conditions: temperature: 24°C (reference temperature: 23 ± 5°C); humidity: 60%. **Figure 6** shows the appearance of the breakage faces by means of a scanning electron microscope (1000×) for cobalt-based alloys.

The fractographic study is correlated with the results obtained in the traction tests and certifies that silicon-alloyed variants exhibit cleavage by break. In the case of fragile fracture, the cracks initiated in the breakage surface propagate sharp without a total deformation of the material, but only in micro-volumes located on the breaking surfaces of the type: $Co_xCr_yMo_z$ with cubic crystalline network, CrSi with cubic crystalline network, and Cr_yMo_zSi with cubic crystalline network (detected by diffractometric X-ray analysis).

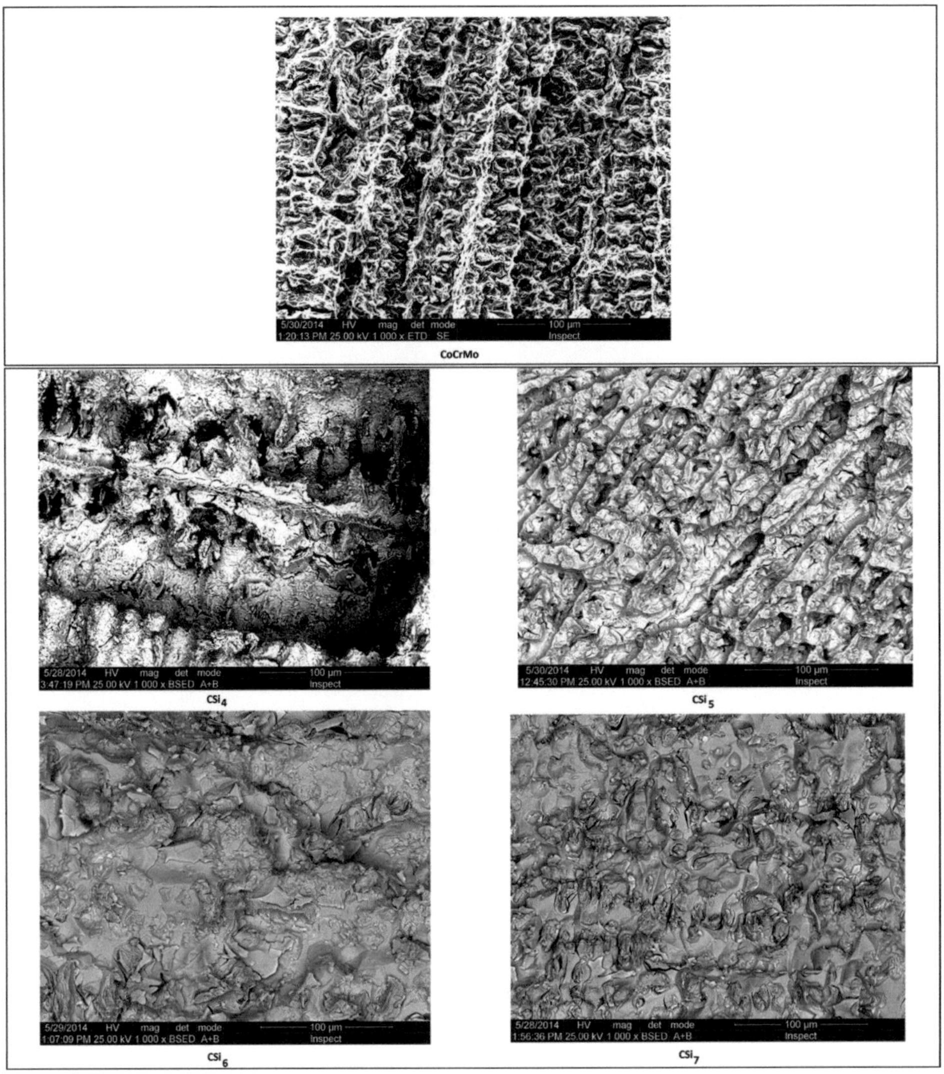

Figure 6. The aspect of cobalt specimen surface observed by electronic scanning microscope (magnification 1000×).

The fragile break produces cuts in a plane approximately perpendicular to the stress plane and has a crystalline aspect (the rupture is initiated on grain boundaries with the aspect of cleavage planes). When the interphase adhesion of the metal matrix is insufficient, there are lamellar strips present on the cracked surfaces that attenuate the quick detachment of the blank space sub-microscopic.

2.8. Characterization by electrochemical impedance spectroscopy (EIS)

Using electrochemical impedance spectroscopy (EIS), information was obtained on the passivation process, which takes place in maintaining the CoCrMo commercial alloy and the new

CSik (k = 4, 5, 6, and 7) in the solution. The advantage of this method (in alternative current) to other electrochemical methods (polarization methods—in DC) is that it can monitor some electrochemical changes over time, being a nondestructive method at the same time.

Electrochemical impedance spectroscopy is a stationary method capable of highlighting relaxation phenomena, whose relaxation times vary in different order of magnitude and allow mediation in the same experiment to achieve a high level of precision.

The CoCrMo alloys subjected to electrochemical characterization in simulated biological environments have a cubic shape with a 10 mm side.

The measurements were made at open-circuit potential, in fresh unpasteurized orange juice. The citric acid (unpasteurized fresh orange juice) was chosen to study electrochemical behavior because it is considered to be one of the five acids that try most often to enter our body, 5–9%. In high quantity, the fresh orange juice can affect the tooth enamel, thinning him, and generating thus the cavities appearance and unpleasant experiences associated to them [12].

The spectrums were recorded in the frequency domain of 10^5–10^{-2} Hz, at a potential in alternative current, with 10 mV amplitude, using a PARTSTAT 4000 potentiostat. Data processing was done with ZSimpWin software, version 3.22. ZSimpWin software uses a varied of electric circuits to correlate numeric the impedance data measured (**Figure 7**).

After each experiment, the impedance data was represented after Bode diagrams (impedance $|Z|$ vs. frequency (f) and phase angle Φ (grade) vs. frequency (f)).

The dependence between phase angle and frequency indicates the fact that may exist one or more time constants, which can be used to determinate the elements values from equivalent circuit.

The advantage of Bode diagrams is that the dates are present for all frequencies measured and the impedance values can be represented on a high interval. It was obtained, after immersion

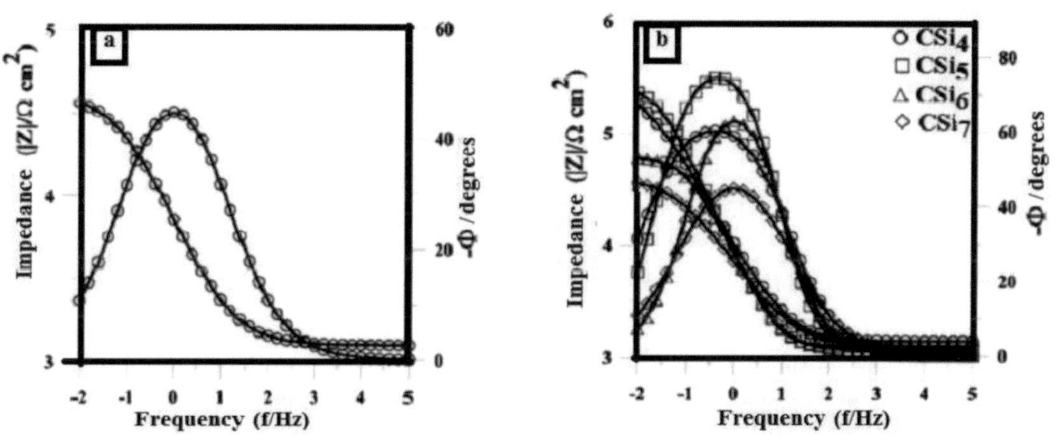

Figure 7. Bode diagrams for studied alloys: (a) commercial alloy from CoCrMo system and (b) CSi_k alloys (k = 4, 5, 6, and 7) experimental alloys, maintaining for 750 s in fresh unpasteurized orange juice.

for 750 s, in fresh unpasteurized orange juice; the impedance spectrums represented like Bode diagrams for four experimental samples from CSi_k alloys (k = 4, 5, 6, and 7). Bode diagrams for alloys from CoCrMo system were present in the **Figure 7a** and **b**.

In Bode representation for commercial alloy from CoCrMo and experimental alloys CSi_k (k = 4, 5, 6, and 7), present only one constant for relaxation time, indicated by a single maximum on variation curve for phase angle with frequency.

The electrochemical cell can be represented by a single equivalent circuit consisting of different combination of resistors, capacitors, and other circuit elements.

The correlation grade of equivalent circuit for obtained experimental data is expressed by χ^2 parameter, which is directly correlated with the relative error of measured current; to one χ^2 value of the order of 10^{-4}, it corresponds to an error of measured value by 2%.

The interpretation of spectrum for all CoCrMo alloys was made by data modeling with an equivalent circuit. For simulation, it used ZSimpWin software. In this equivalent circuit, $R_{sol}(R_1Q_1)$, R_{sol}— solution resistance, R_1—resistance of passive layer (resistance to polarization), and Q_1—capacity of passive layer. In this case, to enlarge the scope of the model, in the place of ideal capacity of passive layer it introduced a constant phase element Q_1. The impedance of this constant phase element is equal with:

$$Q = Z_{CPE} = \frac{1}{Y_o (j\omega)^n} \quad (1)$$

where Q—adjustable parameter (F cm^{-2} s^{n-1}), Y_o—a constant, j—imaginary number ($j_2 = -1$), n—is related to the slope of the lg | Z | versus lg, f—frequency from Bode graphic, and ω is angular frequency.

When the value of n is equal with 1, the constant phase element describes an ideal capacitor (C).

For $0.5 < n < 1$, the constant phase element describes a distribution of relaxation times in the frequency spaces and when n = 0.5, the constant phase element represents a Warburg impedance with diffusion character. When n = 0, the constant phase element describes a resistor.

The χ^2 coefficient values are included between 2×10^{-4} and 5×10^{-4}, which confirm that the chosen equivalent circuit describes well the physic model, adjustment of experimental values being placed in 1–3% error limits. The resistance of solution not varied in the time of samples maintaining in these, the recorded differences for performed measurements varies in ±3 Ω cm^2 limits toward a medium value by 120 Ω cm^2.

The electric parameters of equivalent circuit for CoCrMo alloys maintained 750 s in fresh unpasteurized orange juice are shown in **Table 3**.

From the data present in **Table 3**, it is found that the resistance of passive layer increases once the increase of silicon content, for CSi_4, CSi_5 alloys, which means that, not catalyzes the oxidation process at superficial layer. For samples containing more than 5% Si, CSi_6 and CSi_7, the corrosion resistance decreases gradually.

Alloys	10^{-3} R_1 (Ω cm²)	10^5 Q_1 (S/cm² sⁿ)	n_1
	After 750 s immersion		
CoCrMo	32	3.70	0.79
CSi_4	164	2.40	0.82
CSi_5	295	1.90	0.83
CSi_6	56	3.20	0.80
CSi_7	28	3.70	0.79
	After 3000 s immersion		
CoCrMo	38	3.70	0.79
CSi_4	211	2.30	0.82
CSi_5	435	1.60	0.83
CSi_6	64	3.20	0.80
CSi_7	32	3.70	0.79

Table 3. The electric parameters obtained by the adjustment of experimental data for CoCrMo- and CSi_4-studied alloys in fresh unpasteurized orange juice at varied immersion times.

The electrochemical impedance spectroscopy represented as Bode diagrams for four alloys, CSi_k alloys (k = 4, 5, 6, and 7), and for the commercial alloy (CoCrMo) were obtained as a result of immersion 3000 s in fresh unpasteurized orange juice. The Bode diagrams for CoCrMo alloys are shown in **Figure 8**.

From the shape of Bode spectrometry to variants of silicon alloys, it is found that they exhibit similar electrochemical behavior after 3000 s immersion in fresh unpasteurized orange juice.

The values of the electrical parameters of the equivalent circuit for the studied alloys maintained 3000 s in fresh unpasteurized orange juice are also presented in **Table 3**. In **Figures 7** and **8**, the experimental data are presented as individual points and the continuous line represents the theoretical spectra obtained from the simulation. Long contact between cobalt alloys and in fresh unpasteurized orange juice leads to superficial passivation of the alloys.

From the data presented in **Table 3**, it is found that the resistance of passive layer increases once the increase of silicon content for CSi_5 alloy, which means that it does not catalyzes the oxidation process at superficial layer. It has been noticed that silicon alloying of the CoCrMo alloys are obtained structures and properties that lead to the influence of corrosion resistance.

The modification of properties can take to increased corrosion resistance of alloys in the CoCrMo system. The increase in the corrosion resistance of CSi_4 and CSi_5 alloys at a determined value of the alloying element is explained by the formation of complex structures of the type $Co_{0.8}Cr_{0.2}$ with hexagonal crystalline network and $Co_{0.64}Cr_{0.32}Mo_{0.04}$ with cubic crystalline network identified with using the qualitative phase analysis by diffractometric X-ray investigations. These structures formed on the surface of the experimental alloys, CSi_4 and CSi_5, high pellicle in layer.

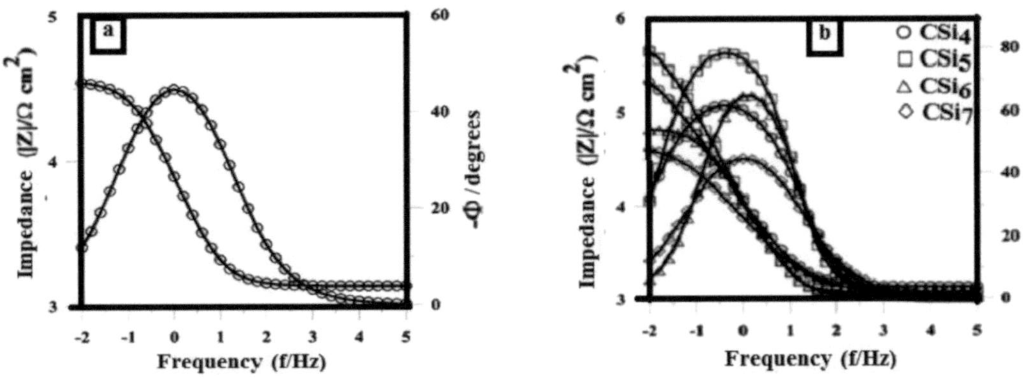

Figure 8. Bode diagrams for studied alloys: (a) commercial alloy from CoCrMo system and (b) CSi_k alloys (k = 4, 5, 6, and 7), maintaining for 3000 s in fresh unpasteurized orange juice.

2.9. Characterization by linear polarization studies

The alloy measurements in the CoCrMo system have been performed in fresh unpasteurized orange juice. The linear polarization curves were indicating in the potential range: −0.8 to +1 V, using a rate of 1 mV/s.

Representation of linear polarization curves in coordinates: current density (j)/potential (E) (**Figure 9a** and **b**) allows to highlight corrosion potentials (E_{cor}) as well as corrosion currents (j_{cor}).

The main parameters of the corrosion process (E_{cor} and j_{cor}) are obtained by processing the linear polarization curves for four CSi_k alloys (k = 4, 5, 6, and 7) and for the CoCrMo commercial alloy are centralized in **Table 4**. Corrosion potentials (E_{cor}) show similar values for the original variants, compared to the value for the CoCrMo commercial alloy. The corrosion current (j_{cor}) is the representative of the degree of damage to the material.

It is found that the density of the corrosion current, representative dimension of the level of degradation of the samples, is in the order of tens of μA/cm² in all the studied cases. According to the experimental results, the value of the corrosion current decreases with the increase in the silicon content to CSi_5, after which it increases by reaching a maximum value for the CSi_7 sample. This "positivity" of cobalt alloy surfaces immersed in fresh unpasteurized orange juice was attributed to the formation of passive layers, most likely oxides (Cr_2O_3 and/or Mo_2O_3) that partially protect the surface of the alloys. The same tendency is observed in the case of the passive current density (j_{pas}).

According to the Stern-Geary equation, the polarization resistance (R_1) is proportional inversely to the density of the corrosion current (j_{cor}) [13]:

$$j_{cor} = B/R_1 \tag{2}$$

where B is the constant that depends on the nature of the material.

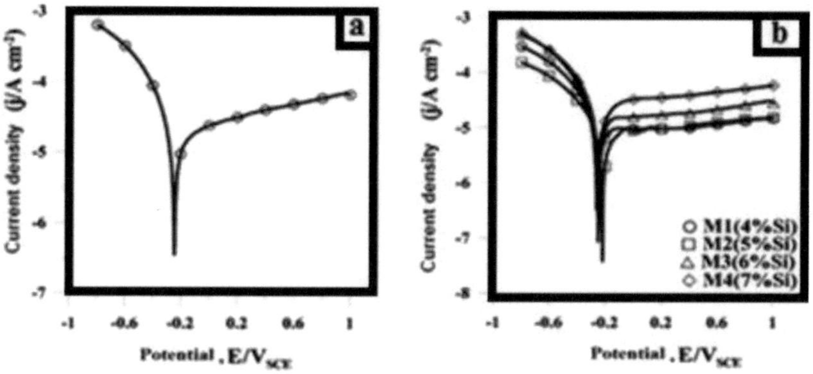

Figure 9. Linear polarization curves for: (a) the CoCrMo commercial alloy and (b) CSi_k alloys (k = 4, 5, 6, and 7) immersed in fresh unpasteurized orange juice, in the potential range: (−0.8 to +1 V), using a rate of 1 mV/s.

Alloy	E_{cor} (mV)	j_{cor} (μA/cm²)	j_{pas} (μA/cm²)
CoCrMo	−245	735	39.1
CSi_4	−247	410	0.9
CSi_5	−217	290	0.9
CSi_6	−243	585	19.4
CSi_7	−257	770	39.6

Table 4. The main parameters of corrosion process for studied alloys in fresh orange juice, unpasteurized.

From the results obtained in the electrochemical impedance spectroscopy tests and those of the linear polarization, we find a good concordance, which is also confirmed by the Stern-Geary equation (**Figure 10**), more exactly the polarization resistance (R_1) is inversely proportional to the current density corrosion (j_{cor}).

2.10. The effect of corrosion on surface layer

In order to confirm the conclusions of the polarization studies and to understand the electrochemical corrosion mechanism of the CoCrMo commercial alloy and CSi_k (k = 4, 5, 6, and 7), the surface microstructure was analyzed by scanning electron microscopy (SEM) with a Vega—Tescan LSH II microscopy.

For the identification of the corrosive effect on surface layer, SEM was used, equipped with back scattered electrons detector (BSE—2D image of surface, best contrast of various phases), at 2000× magnification.

Figure 11 shows the surface morphologies for the CoCrMo commercial alloy and the CSi_k (k = 4, 5, 6, and 7) at the 2000× BSE magnification power, where the impedance spectrometry were recorded after 750 and 3000 s, respectively, immersing the samples in fresh unpasteurized orange juice.

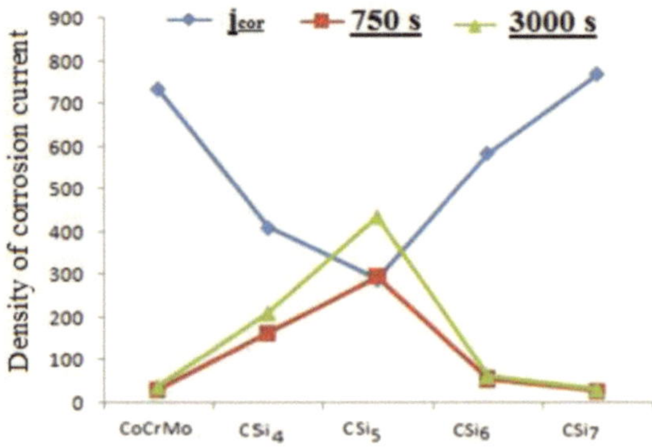

Figure 10. Interface of results (corrosion current density—polarization resistance 750/3000 s) obtained according to the Stern-Geary equation.

The electrochemical sequence used for corrosion resistance testing and analysis of its effects on cobalt alloy surface condition was: linear polarization at −0.8 to +1 V at 1 mV/s. The corrosion points on the alloy surface of 4, 5, and 6% have irregular shape, having different sizes can be observed in the 2000× BSE magnification power. At CSi_7 alloy, the corrosion points have the large dimensions compared to the experimental alloys analyzed.

The SEM micrographs obtained for cobalt alloys indicate that the chemical attack (in fresh unpasteurized orange juice) occurs superficially, uniformly over the entire surface of the alloy, and only locally there are a series of superficial, very small corrosion points, which are randomly distributed on surface [14]. The behavior of the investigated alloys is consistent with the results of potentiodynamic measurements indicating low values for instantaneous corrosion rate (**Table 4**).

2.11. The clinical study on the biocompatibility of the obtained alloys

The central objective of this study was fulfilled by evaluating improvement of the properties of cobalt alloys, used in medical applications, which can resist longtime in human body, as well as knowledge development of new biocompatible structures [15]. If variants of alloys are not subjected to biological acceptance criteria by the animal body, they cannot be placed in the living organism, no matter how appropriate the properties of biomaterials are. The realization of alloy variants should also consider the possibility of the appearance of basic pathophysiological phenomena, which determine their long-term safety (thrombosis, inflammation, infection and/or induction, and neoplasm challenge) [16].

The CoCrMo commercial alloy implanted in subjacent area presents an adipose tissue with canalicular structures and large vessels, separated by membranes of conjunctive tissue with an increased fibroblast population or by islands of conjunctive tissue with similar cell aspect; associated, but and numerous high macrophages, with intracytoplasmic grains (**Figure 12**).

In subcutaneous tissue (**Figure 12a**) after 21 days, can observe a moderate inflammation with a broadcast distribution, rarely nodular, which presents perivascular and perinervous.

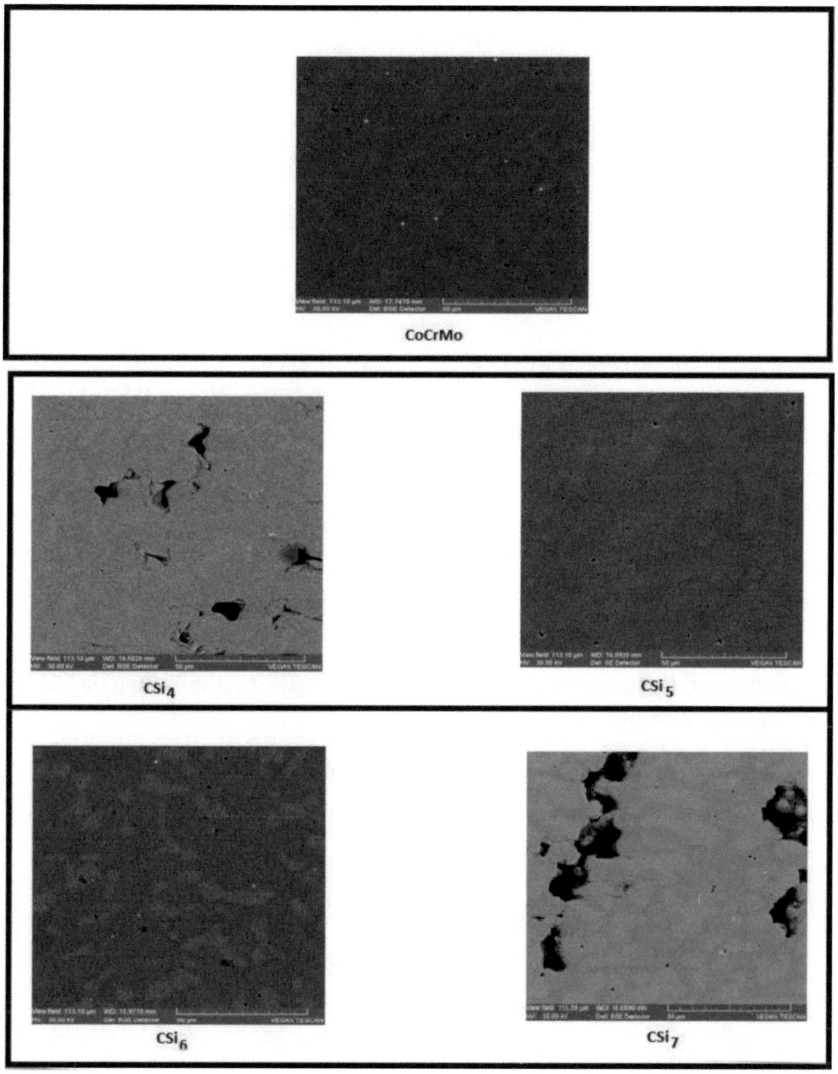

Figure 11. Microstructure of investigated alloys—2000× BSE magnification power.

In subjacent area of implanted commercial alloy (**Figure 12b**), the skin presents normal aspect, with epidermis, without modifications, and a small diminution of the space occupied by dermis, in which are present numerous adipocytes, striated muscle tissue, but and a hypodermis characterized by the presence of a conjunctive tissue richly vascularized, with numerous active fibroblasts.

In lateral area (**Figure 12c**) of implanted commercial alloy (near pill), the skin has an easy papillomavirus epidermis, dermis is superficial and deep, collagenizated intense and hialinizated. It can be observed in this area, near pill exist a hypodermis and muscular tissue normally striated.

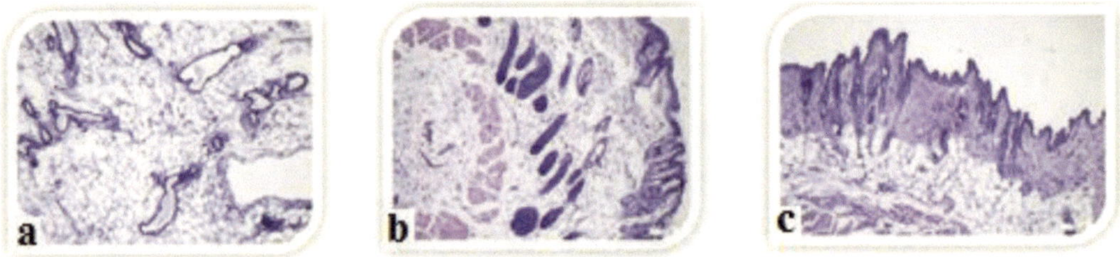

Figure 12. CoCrMo commercial alloy implanted: (a) subjacent area; (b) superjacent area; and (c) lateral area.

CSi_4 alloy implanted in subjacent area (**Figure 13a**), presents a densification of conjunctive tissue with collagen fibers, parallel between then and numerous active fibroblasts, realizing a fibrous band, which delimits the implanted CSi_4 alloy. External to this band, it found a moderate inflammatory infiltrate, numerous macrophages with hemosiderin and numerous congested capillaries.

In superjacent area (**Figure 13b**) of CSi_4 implanted alloy, the skin presents a normal aspect.

At the CSi_4 alloy, it was not taken in lateral area, because the implanted material was surrounded by a compact film, described at subjacent area.

CSi_5 alloy implanted in subjacent area (**Figure 14a**) presents a conjunctive tissue with numerous adiposities and expended venules.

The superjacent area of CSi_5-implanted alloy (**Figure 14b**) presents skin with normal aspect, hypodermis with adiposities and muscular striated tissue.

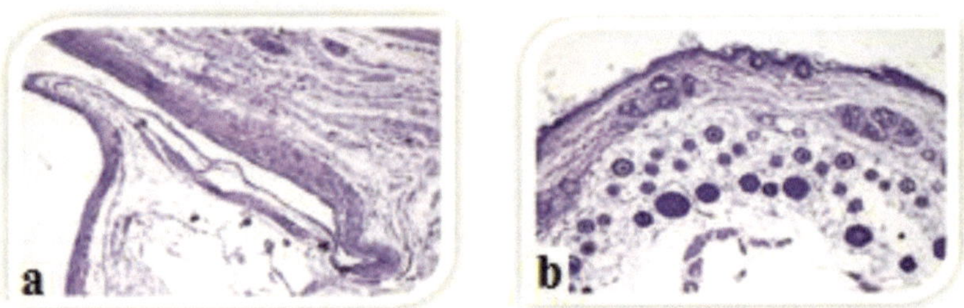

Figure 13. CSi_4 implanted alloy: (a) subjacent area and (b) superjacent area.

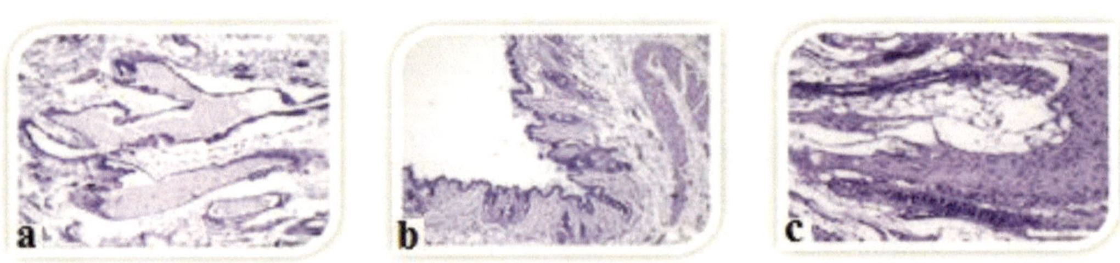

Figure 14. CSi_5 implanted alloy: (a) subjacent area; (b) superjacent area; and (c) lateral area.

http://dx.doi.org/10.5772/intechopen.70886

The skin area is with atrophic epidermis, taking place the diminution of skin axis, and dermis have relatively lax aspect.

The CSi_5 alloy implanted in lateral area (near pill) (**Figure 14c**), highlights a skin with atrophic epidermis. In hypodermis, it found collagenized and hialinizated areas with numerous active fibroblasts. In this area, vascular elements appeared (capillary, arterioles, and venules) with reactive endothelium and numerous lymphocytes in lumen.

In comparison with the CSi_4 alloy, the CSi_5 alloy do not present a conjunctive tissue organized in delimit band of implanted material.

CSi_6 implanted in subjacent area of material (**Figure 15a**) presents a hypodermis with adipose tissue separated by bands and/or collagen islands with numerous fibroblasts. These are infiltrating chronic inflammatory, moderately represented big macrophages with grains.

In the case of CSi_6 alloy, it is found a densification of conjunctive tissue with collagen fibers, parallel between then and numerous active fibroblasts, realizing a fibrous band, which delimits the implanted material. The fibrous band obtained at CSi_6 alloy is thinner than that of CSi alloy.

In the superjacent area of CSi_6-implanted alloy, (**Figure 15b**) can be observed that the skin presents a normal aspect, a hypodermis with aspect similar with that described at subjacent area—densification of conjunctive tissue, under form of fibrous bands, like a wall-organized perimaterial implanted. In the case of this alloy, it is found that the lymphatic vessels are much expended.

In lateral area of CSi_7-implanted alloy (near pill) (**Figure 15c**), the skin presents normal aspect without modifications.

In subjacent area of CSi_7 alloy (**Figure 16a**) have infiltrate inflammatory by massive chronic type, arranged diffusive and/or nodular, perivascullar, and perinervous associating elements by acute type and macrophages.

In superjacent area of CSi_7-implanted alloy (**Figure 16b**), the skin presents a normal aspect, but not a normal hypodermis.

The CSi_7 alloy implanted in lateral (near pill) presents a densification of conjunctive tissue with collagen fibers, parallel between then, and numerous active fibroblasts (**Figure 16c**).

Due to active fibroblasts was made a fibrous band, which delimits implanted material, but is thinner than CSi_4 and CSi_6 original variants.

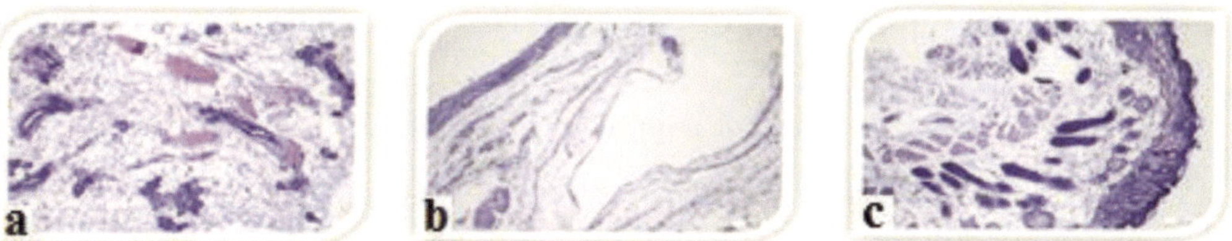

Figure 15. CSi_6 alloy implanted: (a) subjacent area; (b) superjacent area; and (c) lateral area.

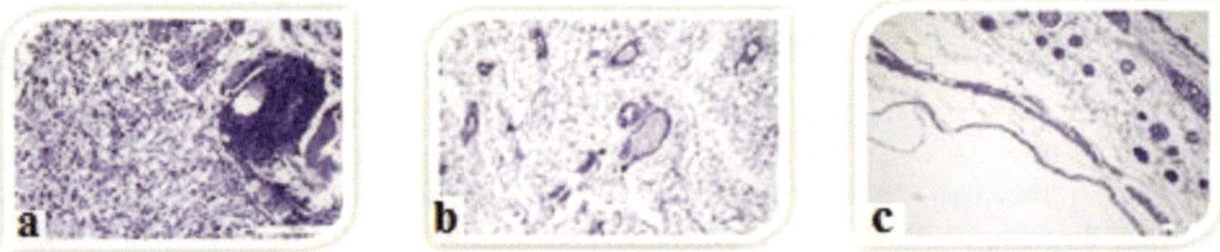

Figure 16. CSi_7 alloy implanted: (a) subjacent area; (b) superjacent area; and (c) lateral area.

The variants of alloys present aspect of biocompatibility detected by microscopic studies, conducted in same conditions like in the case of CoCrMo commercial alloy. The variants of alloys by CoCrMo system do not present major differences in comparison with commercial alloy. The modifications of tissue consist in appearance of one tissue by grain implanted peri-material, which can be interpreted like results of conjunctive organizing and repairing of a consecutive injury.

At variants of CSi_k ($k = 4, 5, 6$, and 7), alloys exist supplementary, adjacent to implanted alloy, a densification of conjunctive tissue, which formed a delimited structure, type fibrous membrane, which is not significant for general state of tested biological organism, because the inflammation is low.

In the biocompatibility study, it is found that no locally histological modifications are recorded, generated by a possible toxicity of biomaterials, based on cobalt, meaning that these type of alloys can be used successfully in medical applications.

3. Conclusions

This scientific research presents the following advantages:

- The increases of chemical homogeneity grade in entire mass for cobalt-base alloy.
- The obtaining of dendrite, biphasic, and uniform structure, which offer an excellent machinability for final products.
- The CoCrMo alloy alloyed with Si, obtained by proposed method, present physical, mechanical, and osteo-integration properties superior than the CoCrMo and CoCr alloys, developed by direct alloying method.
- The ingots obtained by cast, from proposed used alloy, are exactly rigid and nondeformed, even at the small used cavities.
- The remelting installation does not require superheating for melting the alloy, infact it reduces the losses of alloying elements.

The important physical-mechanical characteristics like tensile strength, elasticity modulus, good hardness, and low density ensure the mechanical compatibility of alloy, which is necessary for integration in live tissue and the success for long time of implants.

Both European and National Standards, and also the law requirements from materials domain for medical implanted devices, imposed for biomaterials a high biological compatibility, this being a property influences the behavior of implant in human bodydirectly.

Author details

Petrica Vizureanu[1]*, Mirabela Georgiana Minciuna[1], Gianina Iovan[2] and Simona Stoleriu[2]

*Address all correspondence to: peviz2002@yahoo.com

1 "Gheorghe Asachi" Technical University Iasi, Iaşi, Romania

2 Faculty of Dental Medicine, University of Medicine and Pharmacy "Gr. T. Popa" Iaşi, Iaşi, Romania

References

[1] Balan A, Stoleriu S, Andrian S. Morphological of same commercial dental amalgams. Revista de Chimie. 2015;**66**(2):182-186

[2] Bauer S, Schmuki P, Mark K, Park J. Engineering biocompatible implant surfaces. Materials and Surfaces, Progress in Materials Science. 2013;**58**(3):261-326

[3] Black J. Biological Performance of Materials: Fundamentals of Biocompatibility. 3rd ed. New York: Marcel Dekker Inc; 1999

[4] www.lamet.ro [Internet]

[5] Manual de utilizare—Instalaţie de retopire cu arc, în vid MRF ABJ 900 [Internet]

[6] Geantă V, Voiculescu I, Ştefănoiu R, Chiriţă I. Obtaining and characterization of biocompatible Co-Cr as cast alloys. Key Engineering Materials. 2014;**583**:16-21 Trans Tech Publications, Switzerland

[7] Gupta KP. The Co-Cr-Mo (cobalt-chronium-molybdenum) system. Journal of Phase Equilibria and Diffusion. 2005;**26**(1):87-92

[8] Ghiban B, Borţun CM, Ghiban A, Ghiban N. Cobalt based alloys for dental applications. Key Engineering Materials. 2014;**583**:175-178

[9] Minciună MG, Vizureanu P. Cobalt alloys research used in medical applications. Metalurgia International. 2013;(Special Issue 6):123-126

[10] Vizureanu P, Minciună MG, Achiţei DC, Sandu AV, Hussin K. Mechanical behaviour of CoCrMo alloy with Si content. Applied Mechanics and Materials, Key Engineering Materials. 2015;**754-755**:1017-1022 Switzerland

[11] Minciună MG, Vizureanu P, Achiţei DC, Ghiban N, Sandu AV, Forna NC. Structural characterization of some CoCrMo alloys with medical application. Revista de Chimie (Bucharest). 2014;**65**(3):335

[12] Minciună MG, Vizureanu P, Achiţei DC, Ghiban B, Sandu AV, Mareci D, Bălan A. Electrochemical behavior of CoCrMo and $CoCrMoSi_5$ alloys at different simulated physiological medium. Revista de Chimie (Bucharest). 2014;**65**(10):80-87

[13] Stern M, Geary A. Electrochemical polarization. I. A theoretical analysis of the shape of polarization curves. Journal of the Electrochemical Society. 1957;**104**:56-63

[14] Vizureanu P. Metode și tehnici de cercetare în domeniu. Studiul materialelor prin microscopie optică. Iasi: Universitatea Tehnică, Gheorghe Asachi ed; 2011

[15] Laboratorul Universitar al Bazei Clinice de Învăţământ Medical Stomatologic. www.umfiasi.ro [Internet]

[16] www.cantacuzino.ro/ro/wp-content/uploads/2009/02/catalog-2009.PDF [Internet]

Chapter 3

Synthesis of Cobalt-Based Nanomaterials from Organic Precursors

Maria Smyrnioti and Theophilos Ioannides

Additional information is available at the end of the chapter

http://dx.doi.org/10.5772/intechopen.70947

Abstract

Development of efficient and low-cost methods for the production of cobalt and cobalt oxide nanoparticles is of great interest. Such nanoparticles are typically prepared via transformation of precursors under controlled conditions. In the case of organic precursors, the production of said nanoparticles takes place through thermal decomposition of the organic moiety. The decomposition pathway of the precursor is greatly dependent on the type (i.e. inert, reducing or oxidizing) of the gaseous atmosphere prevailing during heating, as well as on the heating schedule itself. The characteristics of the organic group have also an important influence on the structure of the final material. The goal of the current work is to present a comprehensive review of the research work focusing on the synthesis of cobalt-based nanomaterials from activation of organic precursors.

Keywords: pyrolysis, carboxylates, thermal treatment, cobalt oxides, cobalt, Co_3O_4, nanoparticles, nanomaterials

1. Introduction

Nanostructured materials exhibit novel properties, which notably differ from those of the corresponding bulk solid, due to the small size effect [1]. Transition metal-oxide nanoparticles have attracted considerable interest on account of their applications in various fields, including catalysis, electronics, coatings, gas sensing, electrochemical devices, solar energy conversion and biomedicine [2, 3]. Several methods have been employed for the synthesis of cobalt nanomaterials, including: (i) microemulsion, (ii) solvothermal method, (iii) thermal decomposition of cobalt precursors, (iv) spray pyrolysis, (v) sol-gel process, etc. [1, 4, 5]. Thermal decomposition is a commonly employed method for preparation of nanoscale cobalt or cobalt oxide, due to process simplicity, short reaction time, and, most importantly, low-cost and efficiency. Such nanoparticles are typically prepared via transformation of precursors under controlled conditions. There are many potential precursors, such as, inorganic and organic salts and metal-organic frameworks (MOFs). The decomposition pathway of the precursor depends on the type of the gaseous atmosphere prevailing during heating, on the heating schedule itself, as well as on the characteristics of the organic group. Co-addition of other metal cations toward a mixed metal-organic precursor provides a straightforward approach to the synthesis of mixed metal oxides. Additionally, supported cobalt-based nanomaterials (i.e. cobalt oxide/graphene composites) can be fabricated. The aim of the current work is to present a comprehensive review of the research focusing on synthesis of cobalt-based nanomaterials from activation of organic precursors.

2. Cobalt and cobalt oxide nanoparticles from carboxylates

A major group of organic precursors that have been employed for the synthesis of cobalt and cobalt oxide nanoparticles via a thermal decomposition route is cobalt-carboxylate salts. The "carboxylate" term includes several organic species that contain at least one carboxyl group, such as oxalate, citrate, etc. Methods for synthesis of cobalt-carboxylates include (i) sol-gel [2, 4], (ii) intimate mixing of cobalt ions [6–8], (iii) precipitation [9, 10], (iv) hot oxidation-redox reaction [11], (v) ball milling [12] and (vi) microemulsion [13]. Reaction of cobalt acetate with a carboxylic acid, for example, leads directly to the synthesis of cobalt-carboxylate, when the carboxylic acid in question is stronger than acetic acid: $Co(Ac)_2 + RCOOH \rightarrow Co(RCOO)_2 + 2\ AcH$ (1). This cannot be the case when cobalt nitrate or cobalt chloride is employed, since nitrate and chloride anions are weak conjugate bases of the strong nitric and hydrochloric acids. Nevertheless, many research groups choose to utilize metal-inorganic salts to synthesize cobalt-carboxylate precursors. Thermal treatment of the synthesized carboxylate precursor leads to the removal of the organic species, thus cobalt or cobalt oxide nanoparticles are formed. In the following sections, a review of the synthetic protocols for synthesis of cobalt-based nanomaterials from cobalt-carboxylates is presented.

2.1. Cobalt oxide nanoparticles from oxalate precursors

There are several publications concerning the synthesis of cobalt or cobalt oxide nanoparticles from oxalate precursors. Abdelkader et al. synthesized Co_3O_4 nanoparticles by calcination of cobalt (II) oxalate dihydrate in static air at 500°C for 2 h [14]. Thota et al. employed a sol-gel process to synthesize oxalate precursors and their subsequent thermal treatment to produce Co_3O_4 nanocrystallites. More specifically, cobalt acetate tetrahydrate was dissolved in ethanol at 35–40°C under continuous stirring followed by oxalic acid addition to yield a thick gel. Thermal decomposition of the dried gel in air at 400–600°C for 2 h led to Co_3O_4 [4]. Luisetto et al. have also produced Co_3O_4 by thermal treatment of cobalt oxalate precipitate at 450°C for 2 h in air [2]. de Rivas et al. prepared Co_3O_4 nanoparticles via calcination of oxalate nanorods at 500°C in static air [15]. Cobalt oxalate precursors have been also prepared from cobalt nitrate, cobalt chloride or cobaltous hydroxide carbonate. Shen et al. obtained CoO nanoparticles from the decomposition of corresponding oxalates at 500°C for 4 h in air. Oxalates were prepared via precipitation from a solution of cobalt nitrate, oxalic acid and ethanol [16]. Manteghi et al. synthesized Co_3O_4 nanostructures by thermal decomposition of cobalt oxalate synthesized by precipitation from cobalt nitrate or cobalt chloride and ammonium oxalate in the presence or absence of surfactants [17]. Yuan et al. synthesized Co_3O_4 by annealing cobalt oxalate precursors prepared via hydrothermal and rheological phase reaction methods [18, 19].

2.2. Cobalt oxide nanoparticles from citrate precursors

Manouchehri et al. reported on the synthesis of Co_3O_4 and CoO nanoparticles via solid state ball milling of cobalt acetate and citric acid powders and subsequent calcination of the prepared carboxylate precursors [12]. de Rivas et al. also used ball milling for synthesis of cobalt citrate starting from cobalt carbonate and citric acid [20]. An aqueous sol-gel citrate procedure has been employed by many research groups initiating from cobalt nitrate and citric acid and the obtained precursors were thermally decomposed between 300 and 500°C in static air for 4 h [20–22]. Pudukudy et al. chose a simple solid-state mixing of cobalt nitrate with various modifiers such as citric acid, glucose, sucrose and urea. The components were mixed, pulverized and heated at 80°C for 1 h. The bimodal mesoporous spinel cobalt oxide nanomaterials were prepared by calcination at 300°C for 2 h [23].

2.3. Cobalt and cobalt oxide nanoparticles from malonate precursors

Mohamed et al. synthesized cobalt malonates by precipitation using cobalt chloride and disodium malonate or malonic acid. The precipitates were heated up to 500°C at heating rates between 2 and 20°C min^{-1} in a dynamic atmosphere of N_2, H_2 or air resulting in cobalt oxides and metallic cobalt [10]. Stefanescu et al. synthesized nanosized cobalt oxides unsupported [3] or dispersed in a SiO_2 matrix [24] by thermal decomposition of compounds resulting from a redox reaction of cobalt nitrate and diols (1,2-ethanediol, 1,3-propanediol). The produced caboxylates were oxalate, glyoxylate and malonate. The redox reaction between the NO_3^- ions and -OH groups of each diol took place under controlled heating (90–120°C, 2 h) and the resulting powders were annealed at 350 and 1000°C for 3 h in ambient air.

2.4. Cobalt and cobalt oxide nanoparticles from succinate precursors

Ganguly et al. synthesized cobalt and cobalt oxide nanoparticles from cobalt succinate originated from a reverse micellar method using cobalt nitrate, 1-butanol and CTAB (as the surfactant). The collected product by centrifugation was thermally decomposed under N_2 at 650°C and under air at 500°C for 6 h, yielding pure Co and Co_3O_4, respectively [13].

2.5. Cobalt oxide nanoparticles from tartrate precursors

Bhattacharjee et al. reported the synthesis of Co_3O_4 nanoparticles by thermal decomposition at 400–600°C of cobalt (II)-tartrate complex prepared by a modified sol-gel route employing cobalt nitrate and tartaric acid [25]. Palacios-Hernández et al. prepared CoO and Co_3O_4 nanoparticles by thermal treatment of cobalt tartrate synthesized by precipitation. Cobalt tartrate was obtained by mixing solutions of cobalt nitrate and the disodium salt of L-tartaric acid. The nanoparticles were prepared by thermal decomposition of the metal-organic complex under isothermal conditions at 500°C in the presence of air for 4 h [26].

2.6. Cobalt oxide nanoparticles from other carboxylate precursors

Chen et al. synthesized CoO nanoparticles using cobalt oleate as precursor. Cobalt chloride and sodium oleate were used as initial chemicals and the obtained dried precipitate was dissolved with oleic acid in 1-octadecene. The CoO nanoparticles were produced by heating the mixture up to 320°C and maintaining that temperature for 30 min [27]. Assim et al. presented the synthesis of a series of ethylene glycol-functionalized cobalt (II) carboxylates by an anion exchange reaction of cobalt (II) acetate tetrahydrate with the corresponding acids. Co_3O_4 nanoparticles were prepared by solid state thermal decomposition of $Co(CO_2CH_2(OC_2H_4)_2OMe)_2$ at 300°C, while different decomposition times of 10, 20 and 30 min were applied [28]. Semenov et al. prepared cobalt composites via thermal treatment under isothermal conditions (335°C for 9 h under Ar) of unsaturated cobalt dicarboxylates (hydrogen allylmalonate, acetylenedicarboxylate, itaconate, maleate, glutaconate, cis-muconate) using cobalt carbonate and an unsaturated dicarboxylic acid aqueous solution [9].

3. Mixed metal-oxide nanoparticles from carboxylate precursors

Mixed metal oxides contain two or more metallic ions in proportions that may either be defined by stoichiometry or variable. One way to classify mixed metal oxides is according to their crystalline structure (spinels, perovskites, etc.) [29].

3.1. Spinel mixed metal-oxide nanoparticles from carboxylate precursors

The preparation methods for $CoFe_2O_4$ nanoparticles require specialized techniques to prevent agglomeration. Diodati et al. synthesized various spinel ferrites ($CoFe_2O_4$, $NiFe_2O_4$, $ZnFe_2O_4$ and $MgFe_2O_4$) by a wet-synthesis coprecipitation route starting from iron(III) nitrate, cobalt(II) chloride and oxalic acid in order to prepare the oxalate precursors. The precipitate was centrifuged, dried and calcined at 873 or 1173 K for 5 h in air [30]. Many researchers used the redox method to synthesize the carboxylate precursors (oxalate, malonate, succinate, etc.) initiating from iron(III) nitrate, cobalt(II) nitrate and various diols (ethylene glycol, 1,2-propanediol, 1,3-propanediol and 1,4-butanediol). The obtained precursors were calcined at various temperatures in air to cobalt ferrites [11, 31, 32]. In some cases, the synthesized ferrites were dispersed in a SiO_2 matrix using a modified sol-gel and redox reaction method [31, 33–35]. Amiri et al. synthesized $CoFe_2O_4/SiO_2$ nanocomposites by a procedure described as sol-gel autocombustion method. Iron and cobalt nitrate were dissolved in water and acid solutions (maleic, malonic, trimesic and ascorbic acid) were added separately as well as tetraethoxysilane. The sols were transformed into gel at 40°C for 24 h [36]. Cobalt-zinc ferrite, $Co_{1-x}Zn_xFe_2O_4$ (x varying from 0 to 1) was obtained by thermal decomposition of the carboxylate precursor, starting from a mixture of Fe(III), Co(II) and Zn(II) nitrates and 1,3-propanediol whereas a redox reaction took place. The precursor was decomposed at 300°C and further annealed in air for 3 h at various temperatures from 350 to 1000°C [37, 38].

Mixed cobalt-manganese oxide is another spinel synthesized by thermal decomposition of carboxylate salts. Papadopoulou et al. obtained Co-MnO catalysts from the pyrolytic decomposition of mixed cobalt-manganese fumarate and gluconate salts. The precursor compounds were prepared by mixing aqueous solution of cobalt and manganese acetate with an aqueous solution of D-gluconic acid or an ethanolic solution of fumaric acid. The corresponding mixed metal/metal-oxide nanoparticles were prepared by thermal decomposition of the organic precursors under inert gas flow in the temperature range of 500–700°C. Cobalt-manganese spinel oxide was also prepared via oxidative treatment of the same mixed carboxylates at 500°C for 2 h [6–8]. Faure et al. prepared $Co_xMn_{3-x}O_4$ ($0 \leq x \leq 3$) by controlled decomposition of mixed oxalates, precipitated at room temperature using cobalt nitrate, manganese nitrate and ammonium oxalate. The thermal decomposition of oxalates was carried out near 200°C under O_2/Ar flow and further calcination up to 300°C for 1 h [39].

Wang et al. synthesized $CoAl_2O_4$ nanoparticles via thermal treatment of Co-Al-glycine precursors, initiating from cobalt and aluminum nitrate and glycine. The precursor was treated under air at 200–1000°C for 2 h [40]. Synthesis of $CoCr_2O_4$, a member of the chromite spinel family, was reported by Gingasu et al. via thermal decomposition of tartrate and gluconate precursors obtained by precipitation using cobalt (II) nitrate, chromium (III) nitrate and tartaric acid or δ-gluconolactone [41]. Finally, synthesis of metal cobaltite spinels ($MgCo_2O_4$, $NiCo_2O_4$) by thermal decomposition of coprecipitated mixed oxalate precursors has been reported [42, 43].

3.2. Perovskite mixed metal nanoparticles from carboxylate precursors

Seyfi et al. synthesized unsupported perovskite samples (i.e. $LaCoO_3$) by thermal decomposition of citrate complexes using lanthanum(III) nitrate, cobalt(II) nitrate and citric acid as initial compounds. The obtained viscous gel was dried and calcined at 700°C for 5 h under air atmosphere [44].

3.3. Ilmenite mixed metal-oxide nanoparticles from carboxylate precursors

Gabal et al. prepared $CoTiO_3$ via a coprecipitation method. The mixed oxalate precursor was thermally treated at various temperatures (230–950°C, 30 min and 1000°C, 2 h) and $CoTiO_3$ nanoparticles were obtained [45].

4. Supported cobalt oxide from carboxylate salts

Transition metal oxides have been proposed as materials for lithium-ion batteries (LIBs) and it has been reported that the performance of cobalt oxides is further improved if cobalt oxide nanoparticles are supported on carbon (nanofibers, nanoflakes, etc.) [46]. From this point of view, Guo et al. synthesized Co_3O_4/graphite composites by precipitation of cobalt oxalate on the surface of graphite and pyrolysis of the precipitate [47].

5. Cobalt-based nanomaterials from metal-organic frameworks (MOFs)

A number of recent publications have reported on the synthesis of cobalt-based nanomaterials embedded in carbon, employing metal-organic framework compounds as precursors. Metal-organic frameworks (MOFs) are an emerging class of porous materials constructed from metal-containing nodes and organic linkers [46]. Wang et al. showed that the direct pyrolysis of cobalt nitrate accommodated by impregnation in the pores of ZIF-8 and ZIF-67 yields Co_3O_4 nanoparticles [48]. Ashouri et al. utilized the thermal decomposition of cobalt-terephthalate MOF precursor as a synthetic route for the fabrication of Co_3O_4 nanoparticles [49]. Pei et al. synthesized carbon-supported and Si-doped carbon-supported Co nanomaterials by the thermolysis of Co-MOF-71 at 600°C for 8 h under He flow [50]. Many researchers have utilized MOFs as precursors for the synthesis of cobalt-based nanoparticles embedded in carbon. Zhou et al. prepared nitrogen-doped carbon-embedded Co catalysts by one-step pyrolysis (600–900°C, for 8 h, under N_2) of ZIF-67 with silica as the hard template [51] while, Khan et al. used direct carbonization of ZIF-12 at 950°C for 6 h under Ar atmosphere [52]. Lu et al. fabricated metallic Co nanoparticles embedded in N-doped porous carbon layers by thermal treatment of macroscale Co-MOFs obtained by using 1,3,5-benzenetricarboxylic acid, triethylamine and nonanoic acid, at various temperatures (800, 900, 1000°C) for 1 h under N_2 atmosphere [53]. Qiu et al. obtained cobalt-based nanoparticles coated with a thin carbon shell from Co-MOF-74 via a two-step carbonization where at first the cobalt precursor was heated at 700°C for 2 h under Ar atmosphere to produce cobalt nanoparticles coated with carbon shell and in a second phase the as-synthesized material was calcined either at 500°C for 1 h under CO_2 atmosphere to produce carbon-coated Co_3O_4 nanoparticles or for 30 min under the same conditions to produce Co/CoO nanomaterials. Pure Co_3O_4 nanocrystals were obtained by treating carbon-coated cobalt nanoparticles at 250°C for 2 h under air atmosphere [54].

6. Structure and characterization of cobalt-based nanoparticles

6.1. Structure and characterization of Co_3O_4 nanoparticles

The majority of publications for the synthesis of cobalt-based nanoparticles from thermal decomposition of organic precursors under air have reported on the formation of the spinel

Co_3O_4 phase. The published X-Ray Diffraction (XRD) patterns have reflection peaks that can be perfectly indexed to the cubic phase of Co_3O_4 spinel with the cell parameter α in the range 8.07–8.15 Å and space group Fd3m (**Figure 1**). In most cases, there were no diffraction peaks related to CoO or metallic cobalt (Co^0), except for thermal treatment above 900°C, where Co_3O_4 decomposes to CoO. The average crystallite size, estimated by the Debye-Scherrer equation, varied between 5 and 50 nm, depending on the preparation method and the type of cobalt precursor, the thermal treatment conditions, etc. Structural details of Co_3O_4 have been investigated by Scanning Electron Microscopy (SEM) and Transmission Electron Microscopy (TEM). SEM results indicated the agglomeration of Co_3O_4 nanoparticles and various types of morphologies were observed, that is, matchstick type bars [4, 17], nanorods of bamboo stick type [4], truncated hexahedral shape [18], flocculent-like structures [48] and irregularly shaped aggregates (dense clusters) [21, 22, 26, 49]. TEM results were relatively consistent with XRD results concerning the crystallite size (5–50 nm). Palacios-Hernández et al. reported the synthesis of Co_3O_4 with average grain size of 300 nm and a scarce number of nanoparticles with a diameter of 100 nm and smaller, according to TEM. Most nanoparticles had faceted morphology and high crystalline quality as verified by Selected-Area Electron Diffraction (SAED) analysis and High-Resolution Transmission Electron Microscopy (HRTEM) [26]. Pseudo-hexagonal [15, 17, 18, 21], circular and occasionally rectangular [2] or uniform spherical [12] shaped nanoparticles were observed. In general, powders are homogeneously dispersed with dimensions ranging from 50 to 600 nm, in accordance with SEM results, mentioned above. TEM images further validated that the porous architectures—agglomerations—are constructed from the interconnection among the primary crystalline nanoparticles with size of a few nanometers. Furthermore, the regular diffraction spots displayed from SAED pattern, validated the cubic face-centered structure of Co_3O_4 [2, 4, 18, 25]. HRTEM analysis also indicated the single crystalline nature of Co_3O_4 [18, 25].

The textural properties of cobalt oxides are typically determined by N_2 adsorption-desorption at liquid N_2 temperature. The reported Brunauer–Emmett–Teller (BET) specific surface area and Barrett-Joyner-Halenda (BJH) pore size distribution of Co_3O_4 nanoparticles

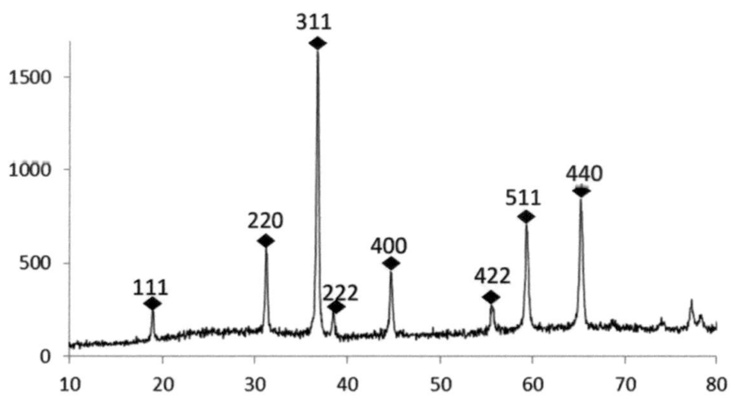

Figure 1. Typical XRD pattern of pure Co_3O_4 [17].

differ significantly from sample to sample. Factors like the synthesis method of cobalt precursor, its nature, the treatment temperature and soak time affect their textural properties. The specific surface areas of Co_3O_4 nanoparticles vary from 7 [23] to 120 $m^2 g^{-1}$ [21], the pore volume ranges between 0.03 [23] and 0.24 $cm^3 g^{-1}$ [20] while average pore size is between 9 [20] and 68 nm [23]. Wang et al. found that all N_2 sorption isotherms of Co_3O_4 and Co-Ce oxide nanoparticles can be assigned to a type IV isotherm, typical for mesoporous materials, showing a hysteresis loop at $P/P_0 = 0.45$ [22]. The mesoporous structure of the nanomaterials was also noted by other researchers [15, 17, 20, 23]. Pudukudy et al. synthesized Co_3O_4 spinels possessing a bimodal mesopore distribution in the presence of modifiers [23]. Manteghi et al. also suggested the presence of a secondary porous structure formed by aggregation of primary nanostructures [17]. Another very useful technique for the characterization of cobalt oxide and its precursors is Fourier Transform Infrared spectroscopy (FTIR) spectroscopy. The appearance of characteristic infrared absorption bands of cobalt precursor and final product obtained from thermal decomposition of the precursor validates their formation. The IR spectrum of cobalt oxide exhibits two major bands at ~570 cm^{-1} (ν_1) and ~670 cm^{-1} (ν_2) (**Figure 2a**). The first band (ν_1) is associated with the Co^{3+} vibration in the octahedral hole and the second band (ν_2) is attributed to the Co^{2+} vibration in the tetrahedral hole, which confirms the formation of the Co_3O_4 spinel [12, 17, 23, 24, 48, 49]. Raman spectroscopy has also been utilized for the confirmation of Co_3O_4 formation. Typically, five Raman bands at 198, 484, 522, 622 and 694 cm^{-1} are visible in the range 100–800 cm^{-1}, which correspond, respectively, to the F^1_{2g}, E_g, F^2_{2g}, F^3_{2g} and A_{1g} modes of crystalline Co_3O_4 (**Figure 2b**). F_{2g} and E_g modes are associated with the vibration of tetrahedral and octahedral sites, whereas the high-frequency band, A_{1g} mode, is linked to the occurrence of octahedral sites [14, 15, 20–22, 25].

The surface chemical status and composition of the nanomaterials can be investigated by means of X-ray photoelectron spectroscopy (XPS). According to XPS results, the Co_{2p} spectrum is decomposed into two distinct peaks and two weak satellites (**Figure 3a**). The two spin orbit

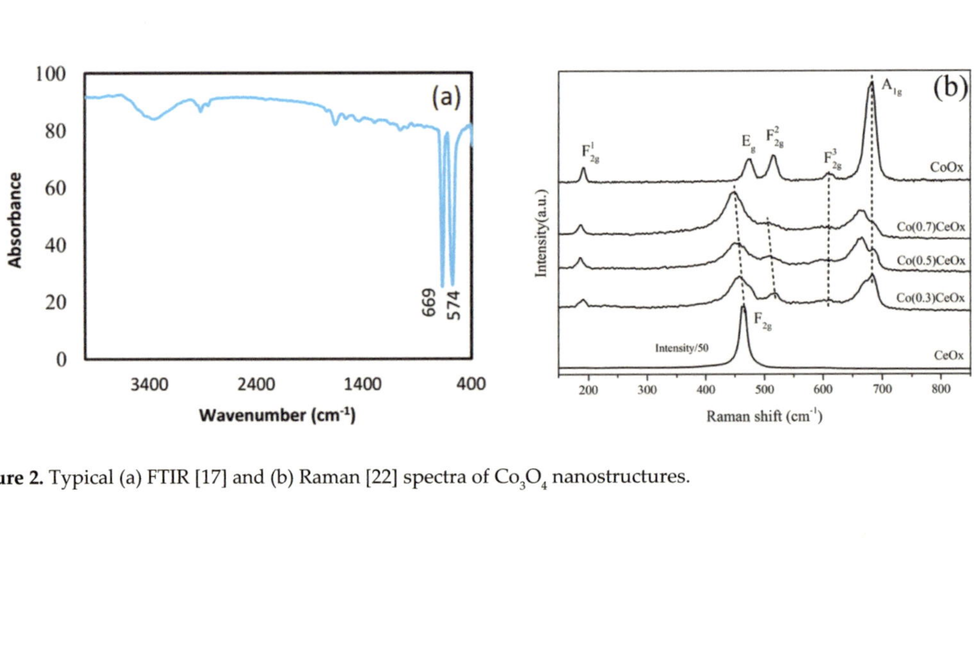

Figure 2. Typical (a) FTIR [17] and (b) Raman [22] spectra of Co_3O_4 nanostructures.

http://dx.doi.org/10.5772/intechopen.70947

doublets assigned to cobalt oxides located at 780.8 and 795.9 eV can correspond to $Co_{2p(3/2)}$ and $Co_{2p(1/2)}$, respectively. The spin orbit doublet of $Co_{2p(3/2)}$ can be deconvoluted into two peaks at 780.8 and 782.3 eV, which are attributed to $Co^{3+}_{2p(3/2)}$ and $Co^{2+}_{2p(3/2)}$ configurations, respectively. The $Co_{2p(1/2)}$ spin orbit doublet can also be deconvoluted into two distinct peaks located at binding energies of 795.8 and 797.4 eV and are assigned to $Co^{3+}_{2p(1/2)}$ and $Co^{2+}_{2p(1/2)}$, respectively. The energy difference of 15.1 eV between $Co_{2p(3/2)}$ and $Co_{2p(1/2)}$ splitting is characteristic of the Co_3O_4 cubic phase. The presence of the two satellite peaks ($Co_{sat.}$) in the vicinity of the two spin orbit doublets, at binding energies of 788.2 and 804.8 eV, further demonstrates the existence of cobalt oxides [22, 55]. The XPS spectrum of O_{1s} presented in **Figure 3b** demonstrates a peak at 530.0 eV attributed to lattice oxygen of Co_3O_4 ($O_{latt.}$, O^{2-}) and another one at 531.0 eV associated with adsorbed oxygen onto surface oxygen vacancies and oxygen-containing groups such as hydroxyl species. The additional peak at 532.1 eV can be attributed to physical and chemical adsorption of water on the surface of defects ($O_{adsorbates}$) and apparently is not always present in O_{1s} spectra [22, 55].

Reducibility and acidity-basicity are important properties affecting material behavior in a specific application. Temperature programmed reduction (TPR) of Co_3O_4 nanoparticles shows two main peaks. The first peak is located at 250–360°C and is assigned to reduction of Co^{3+} to Co^{2+} with the concomitant phase change to CoO and the second peak appears at 350–473°C corresponding to reduction of Co^{2+} to Co^0. Peak temperature deviations in TPR might be due to structural variations of Co_3O_4 nanoparticles (particle size and faceting) and TPR protocols [2, 14, 20–22]. The acidic properties of Co_3O_4 nanoparticles have been studied by NH_3-TPD. Abdelkader et al. performed NH_3-TPD on Co_3O_4 pre-reduced at 400°C for 1 h and reported that the TPD profile contained two peaks, at 104 and 240°C, indicating the presence of two acidic sites [14]. de Rivas et al. carried out NH_3-TPD experiments on Co_3O_4 nanomaterials pretreated in a O_2/He stream at 500°C and observed two ammonia desorption peaks at 190–200°C and 300–305°C as well as a shoulder at 250°C. They concluded that the peak at low

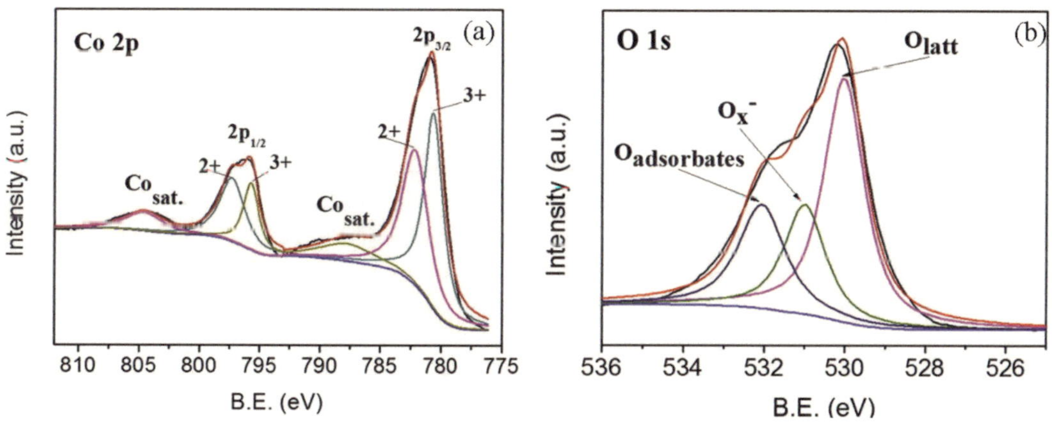

Figure 3. XPS spectra of (a) Co_{2p} region and (b) O_{1s} region [55].

temperature could be associated with weak acidic sites and the peak at high temperature with strong acidic sites [20]. In addition to physicochemical characterization of cobalt nanoparticles, some researchers have studied the optical, electrical and magnetic characteristics of the synthesized nanomaterials by means of UV-Vis-NIR Spectrometry [4], SQUID [4], cyclic voltammetry (CV) [17, 18] and electrochemical impedance spectroscopy (EIS) [17].

6.2. Structure and characterization of CoO nanoparticles

As already mentioned above, there is scarce bibliography concerning the preparation of CoO nanoparticles by thermal decomposition of cobalt-organic precursors. XRD results reported by Bartůněk et al. [5] and Manouchehri et al. [12] showed the presence of a face-centered cubic phase with cell parameter $\alpha = 4.26$ Å and space group F_3m [12]. In the FTIR spectrum, the appearance of a broad band at 450 cm^{-1} verified the formation of cubic CoO. According to TEM the as-prepared CoO consisted of two types of structures, small nanoparticles and nanorods [12]. Shen et al. claimed that the CoO nanoparticles exhibited quasi-spherical shape and their measured—from TEM micrographs—average particle size was 18 nm [16]. Chen et al. reported on flower-like CoO nanoparticles with an average particle size of 50 nm [27]. In some cases, synthesis of Co_3O_4 nanoparticles was accompanied by the formation of CoO or metallic cobalt as impurities. Pursuant to XRD results face-centered cubic (space group: F_3m) [23], monoclinic (space group: C2/m) [23] and tetragonal [9] crystal systems were reported.

6.3. Structure and characterization of metallic Co nanoparticles

Ganguly et al. produced pure metallic Co nanoparticles confirmed by XRD results with average size, as measured by TEM, between 10 and 40 nm [13]. Co composites were also synthesized by Semenov et al. containing, according to XRD data, metallic Co (hexagonal symmetry, unit cell parameters: $a = 2.506$ Å, $c = 4.071$ Å) and CoO as admixtures in the main phase of Co_3O_4 [9].

6.4. Structure and characterization of mixed metal-oxide nanoparticles

Dippong et al. synthesized $CoFe_2O_4$ nanoparticles and concluded that single phase cubic spinel $CoFe_2O_4$ can be produced at all temperatures initiating from 1,3-propanediol whereas, when initiating from 1,2-ethanediol and 1,2-propanediol elevated temperatures are required. The degree of $CoFe_2O_4$ crystallization increased with the increase of calcination temperature. The appearance of the intense, symmetric band at 560–585 cm^{-1} in the FTIR spectra, characteristic of the $CoFe_2O_4$ phase, confirmed its formation, while, TEM results revealed—in accordance with XRD—the presence of $CoFe_2O_4$ as nearly spherical nanoparticles with dimensions varying with calcination temperature between 8 and 20 nm [11]. Another publication from the same group focused on the influence of Co/Fe ratio on the oxide phases in $CoFe_2O_4$ nanoparticles and proved the dependence of nanoparticle size on sample composition (Fe and Co content) and calcination temperature. The diameter of the as-synthesized nanoparticles ranged between 12 and 88 nm and according to TEM micrographs, the spherical nanocrystallites showed an agglomerated conformation [56]. Similar work has been conducted by Stefanescu et al., who concluded that the average particle size of the nanoparticles estimated from XRD varied in the range 10–19 nm [32]. Diodati et al. synthesized $CoFe_2O_4$ and carried out in-situ temperature-dependent XRD of the oxalate precursors and showed that the XRD patterns of the spinel phase began to be visible at 773–873 K whereas they became sharp and distinct at 1073–1173 K. The crystallite size calculated by means of XRD and TEM was 23 nm (calcination at 873 K) and 185 nm (calcination at 1173 K). Furthermore, TEM micrographs showed that cobalt ferrite nanoparticles retained their identity and size but they had a tendency to form agglomerates. According to XPS and Mössbauer results, the as-synthesized compounds contained surface iron in oxidation state (III) and not in (II). The TPR result showed that the reduction process took place at 700–1173 K and resulted in the reduction of Co(II) and Fe(III) cations to a metallic state [30]. In addition, some researchers have studied the synthesis of $CoFe_2O_4$ embedded in SiO_2 and according to XRD results the formation of $CoFe_2O_4$ spinel phase is affected by various factors, that is, annealing temperature, percentage of silica matrix, the precursor type, etc. For instance, Dippong et al. have shown that the nanocrystallite size and crystallinity increased with increase of $CoFe_2O_4$ and decrease of SiO_2 content [34], with the use of longer chain precursors [31] and the increase of the annealing temperature [36]. The diameter of $CoFe_2O_4$ nanoparticles, calculated by means of XRD and TEM, ranged between 8 and 32 nm depending on the parameters mentioned above and according to SEM images they were spherical and formed irregular agglomerations [31, 34, 35]. Amiri et al. synthesized $CoFe_2O_4$ nanoparticles dispersed in a SiO_2 matrix by thermal decomposition of iron-cobalt-carboxylates (maleate, malonate, trimesate and ascorbate) and according to XRD patterns, the thermal decomposition of iron-cobalt trimesate precursor led to the formation of pure cubic spinel $CoFe_2O_4$ with a crystallite size of 7 nm. The formation of the $CoFe_2O_4$ phase was also confirmed by FTIR spectra due to the appearance of the characteristic bands at 469 and 694 cm^{-1}. FE-SEM analysis showed irregularly shaped nanoparticles with an average size of 36 nm (Fe-Co-malonate), 38 nm (Fe-Co-ascorbate) and 27 nm spherical nanoparticles with (Fe-Co-trimesate) and 37 nm (Fe-Co-maleate). All nanoparticles formed agglomerations according to SEM images [36]. Another nanomaterial prepared using thermal decomposition of the corresponding carboxylate precursors was $Co_{1-x}Zn_xFe_2O_4$. Muntean et al. concluded by means of XRD that the as-synthesized nanomaterials consisted of the cubic spinel phase $Co_{1-x}Zn_xFe_2O_4$ and impurities ($CoFe_2O_4$, $ZnFe_2O_4$) except for the thermal decomposition at 1000°C, where $Co_{0.5}Zn_{0.5}Fe_2O_4$ ferrite was obtained as a single, well crystallized phase. The increase in annealing temperature increased the crystallinity and the crystallite size of the nanoparticles. The formation of mixed spinel structure was also confirmed by FTIR spectra due to the presence of characteristic bands at 400 and 600 cm^{-1} which indicate the anion-cation interaction in octahedral and tetrahedral voids. The average crystallite size varied between 11 and 81 nm depending on the annealing temperature [37]. Similar experiments performed by Stefanescu et al. led to particles with average crystallite size in the range 7–33 nm due to the lower annealing temperature (up to 600°C instead of 1000°C in Ref. [37]) [38].

Mixed cobalt-manganese oxides were investigated by Papadopoulou et al. [6–8]. They employed in-situ XRD to identify the evolving crystalline phases at various stages of pyrolysis of cobalt-manganese gluconate or fumarate precursors and concluded the formation of MnO and metallic Co phases at 550°C due to the thermal decomposition of the fumarate precursor, whereas the decomposition of gluconate precursor above 200°C did not lead to the formation of crystalline phases before 650°C. Same results were obtained by XRD measurements over passivated samples originated from fumarate salts. Thus, the thermal decomposition of both fumarate and gluconate precursors led to the formation of reduced Co-MnO phases in a single step, due to a carbothermal reaction that took place between residual carbon and Co^{2+} ions. Additionally, the specific surface area of all samples obtained from fumarate precursor ranged between 200 and 220 $m^2\ g^{-1}$ regardless of the pyrolysis temperature, due to the presence of residual carbon in the materials which stabilized their porous structure and prevented extensive sintering of MnO and Co crystallites. Furthermore, the same research group synthesized Co-Mn spinel oxide via oxidative treatment of fumarate precursors and the calcined samples had considerably smaller surface area than pyrolyzed ones in the range of 21–35 $m^2\ g^{-1}$ [6–8]. Faure et al. synthesized $Co_xMn_{3-x}O_4$ ($0 \leq x \leq 3$) by thermal decomposition of oxalate precursor and concluded that the oxides with $x < 0.9$ are amorphous, while for $x > 0.9$ all the nanomaterials had a cubic spinel structure. According to TEM and XRD results, the crystallite size ranged between 6 ($x = 1.6$) and 16 ($x = 3$) nm. Regarding textural properties, the BET surface area and the pore volume were the highest for $x = 2$ ($S_{BET} = 270\ m^2\ g^{-1}$, pore volume = 0.48 $cm^3\ g^{-1}$) whereas further increase of cobalt content was associated with a decrease of surface area and pore volume [39].

The structure and physicochemical characteristics of cobalt aluminate ($CoAl_2O_4$) nanomaterials were investigated by means of XRD, FTIR, TEM and SEM. Wang et al. synthesized $Co^{II}Co^{III}{}_xAl_{2-x}O_4$ ($x = 0$–2) and reported that nanomaterials had spinel cubic structure (Fd3m) and that their crystallite size ranged between 9 and 51 nm depending on Co/Al ratio and pyrolysis temperature. Their formation was further validated by FTIR spectroscopy from the appearance of the characteristic vibrational bands of $CoAl_2O_4$ (572 and 668 cm^{-1}). According to SEM images the spinel samples were agglomerated in large particles with irregular shape. The BET surface area was 22–69 $m^2\ g^{-1}$ depending on annealing temperature [40]. Gingasu et al. investigated the physicochemical properties of $CoCr_2O_4$ nanomaterials synthesized by thermal decomposition of tartrate and gluconate precursors. The reflections of XRD patterns could be indexed to the face-centered cubic cell with the Fd3m space group and the average crystallite size obtained from tartrate and gluconate precursors was 14–18 and 21 nm, respectively. According to SEM and TEM micrographs, the particle size of $CoCr_2O_4$ obtained from tartrate was in the range 13–35 nm (depending on annealing temperature), while $CoCr_2O_4$ obtained from gluconate was in the range 16–28 nm showing in each case well-faceted pyramidal or bipyramidal geometry. $CoCr_2O_4$ from gluconate precursor formed spherical aggregates with an average size 200–400 nm. The formation of cobalt chromite nanomaterials was confirmed by the presence of sharp and intense bands in the range 615–640 and 500–530 cm^{-1} in FTIR spectra, which are characteristic of Cr-O bonds. According to Raman spectra, the spinel type $CoCr_2O_4$ has a cubic symmetry where the Co^{2+} and Cr^{3+} ions occupy the tetrahedral and octahedral sites, respectively. Furthermore, the nitrogen adsorption isotherms were of IV type with a H3 hysteresis loop, characteristic for mesoporous materials. The BET surface area and pore parameters for tartrate-derived $CoCr_2O_4$ were S_{BET} = 15–51 $m^2\ g^{-1}$, pore volume = 0.11–0.24 $cm^3\ g^{-1}$ and average pore diameter = 21–32 nm, whereas for gluconate-derived $CoCr_2O_4$ were 17–42 $m^2\ g^{-1}$, 0.15–0.21 $cm^3\ g^{-1}$ and 21–34 nm, respectively [41]. Darbar et al. synthesized $MgCo_2O_4$ by oxalate decomposition and XRD patterns were indexed to the standard cubic structure with Fd3m space group. The SEM images revealed the agglomeration of $MgCo_2O_4$ nanoparticles and the BET surface area was measured in the range 4–22 $m^2\ g^{-1}$ depending on calcination temperature [42]. Kim et al. produced $NiCo_2O_4$ aggregates by oxalate decomposition and the XRD pattern confirmed the formation of cubic spinel $NiCo_2O_4$ phase with minor additional impurities (NiO). SEM micrographs showed the presence of various morphologies, such as micro flowers and rod shapes and according to TEM images each microstructure consisted of aggregates of spherical metal-oxide nanoparticles with diameter between 20 and 100 nm [43]. Seyfi et al. reported on the synthesis of $LaCoO_3$ perovskite catalysts with specific surface area of 5 $m^2\ g^{-1}$ [44]. On the other hand, Gabal et al. produced purely trigonal ilmenite $CoTiO_3$ by thermal treatment of cobalt oxalate-TiO_2 precursor. They concluded that at high temperature (1000°C), $CoTiO_3$ with rhombohedral symmetry (space group R-3 (148)) was formed. According to TEM images, the nanomaterial showed weak agglomeration with heterogeneous morphology in both shape and dimensions and exhibited a larger grain size, 300 nm, compared to that estimated from Scherrer's equation, 209 nm, due to the agglomeration nature of the nanoparticles as observed from the TEM micrograph. In addition, the specific surface area of the sample was 30 $m^2\ g^{-1}$ and the adsorption isotherm was classified to type II, characteristic for macroporous adsorbents and had very small hysteresis loop [45].

6.5. Structure and characterization of supported cobalt oxide nanoparticles

The results from XRD analysis of cobalt compounds supported on carbonaceous carriers differ depending on the nature of the compound (Co_3O_4, CoO, Co) but in all cases reflections corresponding to the lattice planes of graphitic materials have been observed. The reflections due to cobalt/cobalt oxide nanoparticles were similar to those mentioned above for unsupported materials or in some cases they were not detected at all. It is obvious that an increase in the carbon content leads to an increase of the graphitic diffraction peaks and simultaneous decrease of cobalt/cobalt oxide peaks. Additionally, with increasing calcination temperature, the diffraction peaks for cobalt/cobalt oxide became sharper in contrast to those for carbon, indicating that the crystallinity of cobalt-based nanoparticles increased and graphite content decreased due to its partial oxidation [47]. According to SEM and TEM micrographs, cobalt-based nanoparticles were spherical, coated with carbon and dispersed in the carbon matrix [54], uniformly distributed on carbon layered structures with particle size of 5–30 nm [53], supported on carbon with random size distribution in the range from 1 to 32 nm [50], embedded in carbon with average particle size of 8–10 nm [51] and encapsulated within carbon nanotubes with an average particle size of 50 nm [52]. N_2 adsorption-desorption isotherms of the as-synthesized nanocomposites exhibited the type IV isotherm with a H3-type hysteresis loop, demonstrating the mesoporous nature of the materials, while the textural properties of the materials varied widely (S_{BET} = 76–365 m^2 g^{-1}, pore volume = 0.15–0.96 cm^3 g^{-1}, average pore size = 1.5–20 nm) [50, 52–54]. However, Zhou et al. observed a N_2 adsorption isotherm of type V with a H3-type hysteresis loop, while the specific surface area ranged between 160 and 375 m^2 g^{-1}, the pore volume was 0.58–0.98 cm^3 g^{-1} and the average pore size varied from 4.8 to 6.9 nm depending on the calcination temperature [51]. In many publications, electrical and magnetic properties have been investigated due to the potential use of the as-synthesized materials in Li-ion batteries and as electrocatalysts [47, 52–54].

7. Applications of cobalt-based nanoparticles

The structure and physicochemical characteristics of cobalt-based nanomaterials make them suitable candidates as catalysts for various processes, that is, alcohol reforming, Fischer-Tropsch, oxidation of organic compounds and CO, as oxygen electrocatalysts and as electrode materials in lithium-ion batteries.

There has been a significant effort to decrease the dependence on non-renewable fossil fuels and move toward sustainable energy carriers, such as hydrogen derived from renewable sources. H_2 is considered to be the energy carrier of the future and the use of liquid bio-fuels, such as ethanol and methanol, as hydrogen carriers is an attractive option. Thus, H_2 production from alcohol steam reforming has been of great interest and has been extensively investigated. Since alcohol reforming is a catalytic process, various catalytic systems have been tested aiming at optimization of their efficiency. Among them, cobalt-based catalysts are promising candidates. Papadopoulou et al. prepared Co-MnO catalysts by pyrolytic decomposition of the corresponding fumarate and gluconate salts and examined them in methanol and ethanol reforming. They concluded that catalysts with the highest cobalt loading were the most active both in methanol and ethanol reforming. The utilization of fumarate-derived Co-MnO catalysts led to complete methanol conversion at 375–400°C (high-temperature methanol reforming compared to copper-based catalysts), while complete ethanol conversion was achieved at 480°C (low-temperature ethanol reforming compared to noble-metal catalysts). Thus, reforming of both alcohols took place under comparable conditions at temperatures in the range of 400–450°C. The advantages of the proposed method of catalyst preparation were the formation of the reduced active state of the catalysts in a single step and the existence of residual carbon, which hindered sintering and excessive particle growth under synthesis and reaction conditions [6–8]. Abdelkader et al. studied the catalytic activity of Co_3O_4, Fe_2O_3 and corresponding mixed Co-Fe in the steam reforming of ethanol and concluded that the mixed Co-Fe sample exhibited higher H_2 yield, greater selectivity to CO and CO_2 and reduced by-product formation compared to pure Co_3O_4, Fe_2O_3 and physical Co-Fe mixture [14].

Rechargeable lithium-ion batteries (LIBs) are the most utilized energy storage devices for portable electronics and one of the most promising electrode materials are cobalt-based nanomaterials in place of carbon. Novel porous Co_3O_4 architectures [18, 19], $MgCo_2O_4$ nanomaterials [42], Co_3O_4/graphite composites [47], cobalt-based nanoparticles embedded in nitrogen-doped nanotubes [52] and carbon-coated Co_3O_4 nanoparticles [54] have been studied regarding their electrochemical performance. Briefly, Yuan et al. reported on the performance of Co_3O_4 polyhedral architectures and showed that they exhibited high discharge capacities for many discharge/charge cycles, due to their porous structure [18]. Yuan et al. synthesized nanosized Co_3O_4 and concluded that the advantage of this material over carbon was due to its higher capacity per unit volume (7.5 times in comparison to carbon) and that the particle size affected its electrochemical properties (the optimum average particle size was 37 nm) [19]. Guo et al. studied the electrochemical properties of Co_3O_4/graphite composites and showed that the reversible capacity increased with decrease in graphite content and increase in the calcination temperature, while cycling stability decreased dramatically with decrease in graphitic content [47]. Khan et al. reported on the excellent performance of cobalt oxide nanoparticles embedded in nitrogen-doped carbon nanotubes, which is attributed to the nitrogen doping of carbon nanotubes, the strong interaction between the encapsulated cobalt oxide nanoparticles with the carbon nanotubes, the porosity and the specific surface area of the nanomaterial [52]. In addition, Qiu et al. investigated the electrochemical performance of carbon-coated Co_3O_4 nanoparticles in comparison to bare Co_3O_4 electrode and concluded that the superior electrode performance of the first was attributed to better dispersion and to the thin carbon shell coating of the nanoparticles on the electrode surface [54].

Lu et al. showed that the utilization of Co-MOFs for the fabrication of metallic Co nanoparticles embedded in nitrogen-doped porous carbon layers led to the production of an efficient electrocatalyst with bifunctional activities toward oxygen reduction and evolution reactions (ORR, OER) in alkaline media due to nitrogen doping and embedded cobalt nanoparticles in the carbon structure [53].

Concerning environmental catalysis, the strict environmental legislation for pollutant emissions and the high-cost of noble-metal catalysts has shifted interest toward the production of effective transition metal-oxide catalysts. Cobalt-based nanomaterials, such as nanocrystalline Co_3O_4, Co-Ce composite oxide, $CoFe_2O_4$ and Co-Mn oxide spinels have been synthesized by thermal decomposition of organic precursors and studied in various pollutant abatement processes. de Rivas et al. have examined nanocrystalline Co_3O_4 catalysts in the gas-phase oxidation of chlorinated short chain alkanes and reported on their enhanced catalytic behavior as they exhibited high conversion to deep oxidation products (CO_2, HCl, Cl_2) at low temperatures with excellent selectivity to CO_2 and resistance to deactivation [15, 20]. Similar experiments have been conducted by Wang et al. using Co-Ce catalysts [22]. Liu et al. have demonstrated the excellent activity and stability of Co_3O_4-based catalysts for propane catalytic combustion [21]. Faure et al. synthesized Co-Mn-oxide spinel catalysts and concluded that their high activity for CO oxidation was correlated to both their surface area and cobalt concentration. Among the synthesized nanomaterials, $Co_{2,3}Mn_{0,7}O_4$ exhibited higher activity than cobalt oxide catalysts [39]. Co_3O_4 nanoparticles [48] and $LaCoO_3$ nano-perovskite [44] catalysts were also investigated for CO oxidation reaction and exhibited good structural and chemical stability and high activity. Additionally, Ashouri et al. employed MOF-derived Co_3O_4 for the catalytic oxidation and epoxidation of olefins and reported that the catalyst showed good catalytic stability and reusability [49]. Diodati et al. synthesized and used $CoFe_2O_4$ in the catalytic oxidation of methane and concluded that the catalytic reaction took place at lower temperature (by about 300 K) in comparison to the uncatalyzed reaction while the cobalt ferrite catalyst exhibited good stability [30]. Gingasu et al. synthesized cobalt chromite catalysts via thermal decomposition of tartrate and gluconate precursors and reported that gluconate-derived $CoCr_2O_4$ was the best catalyst for total oxidation of methane exhibiting high activity and CO_2 selectivity [41]. Co nanoparticles embedded in mesoporous nitrogen-doped carbon were investigated by Zhou et al. for the reductive amination of carbonyl compounds with nitro compounds by transfer hydrogenation with formic acid and found them to be active and selective with excellent yields for secondary amines [51]. Another reported application of cobalt-based nanoparticles is in water treatment for the removal of dye pollutants. Advanced oxidation processes and especially Fenton reactions are considered to be among the most feasible pollutant degradation technologies [16, 55].

One of the most significant applications of cobalt nanomaterials in catalysis, is in Fischer-Tropsch synthesis. The Fischer-Tropsch process targets the production of synthetic liquid fuels using coal, natural gas and biomass and hydrogen derived from renewable electricity as feedstocks. Cobalt-based catalysts have been employed to a great extent since they have high selectivity for long chain hydrocarbons and low selectivity for the water-gas shift reaction and they are cost-effective. The reducibility, cobalt dispersion and loading and nanoparticle structure are some of the parameters that affect the catalytic activity and selectivity. Pei et al. synthesized fully reduced carbon-supported cobalt catalysts by pyrolysis of Co-MOF-71 precursor and showed that the as-prepared catalysts had extremely high Co site density, high activity and selectivity with diesel fuels being the main products and high C_{5+} space-time yields in comparison to conventional cobalt catalysts [50].

8. Conclusions

Nanostructured cobalt-based materials have attracted considerable interest due to their multiple applications in various fields, like catalysis, electronics, electrochemical devices, etc. Among the synthesized cobalt nanomaterials, Co_3O_4 is by far the main nanostructure been reported in the literature, whereas fewer studies have been made regarding CoO, metallic Co, mixed metal-oxide nanoparticles or other types of cobalt-based nanomaterials. A great number of methods have been employed for their synthesis with the thermal decomposition of organic precursors bearing noteworthy advantages as, for instance, simplicity, short processing time and cost-effectiveness. Carboxylate salts (e.g. oxalate, citrate, etc.) as well as metal-organic frameworks (MOFs) are utilized as organic precursors for the production of cobalt oxides, metallic cobalt, mixed cobalt oxides and supported cobalt-based nanomaterials. The nature of organic precursor, the precursor synthesis method, the cobalt content, the combination with other elements, the treatment schedule and temperature and the type of the gaseous atmosphere (oxidizing, inert, reducing) prevailing during heating are some of the factors that affect the physical and chemical properties of the final composite. The relevant nanomaterials have been extensively characterized and analyzed using various techniques (XRD, FTIR, SEM, TEM, XPS, Raman, etc.) and accounts have been reported on their novel properties in comparison to bulk solids. A non-exhaustive list of potential applications of such cobalt-based nanomaterials includes catalysts for Fischer-Tropsch process, alcohol reforming, oxidation of organic compounds and CO, oxygen electrocatalysts and anode materials for Li-ion batteries.

Author details

Maria Smyrnioti[1,2] and Theophilos Ioannides[1*]

*Address all correspondence to: theo@iceht.forth.gr

1 Foundation for Research & Technology – Hellas, Institute of Chemical Engineering Sciences (FORTH/ICE-HT), Patras, Greece

2 Department of Chemistry, University of Patras, Greece

References

[1] Thangavelu K, Parameswari K, Kuppusamy K, Haldorai Y. A simple and facile method to synthesize Co_3O_4 nanoparticles from metal benzoate dihydrazinate complex as a precursor. Materials Leters. 2011;**65**:1482-1484

[2] Luisetto I, Pepe F, Bemporad E. Preparation and characterization of nano cobalt oxide. Journal of Nanoparticle Research. 2008;**10**:59-67

[3] Stefanescu O, Davidescu C, Muntean C. Preparation and characterization of cobalt oxides nanoparticles starting from Co(II) carboxylate precursors. Journal of Optoelectronics and Advanced Materials. 2015;**17**:991-996

[4] Thota S, Kumar A, Kumar J. Optical, electrical and magnetic properties of Co_3O_4 nanocrystallites obtained by thermal decomposition of sol-gel derived oxalates. Materials Science and Engineering B. 2009;**164**:30-37

[5] Bartůněk V, Huber Š, Sedmidubský D, Sofer Z, Šimek P, Jankovský O. CoO and Co_3O_4 nanoparticles with a tunable particle size. Ceramics International. 2014;**40**:12591-12595

[6] Papadopoulou E, Delimaris D, Denis A, Machocki A, Ioannides T. Alcohol reforming on cobalt-based catalysts prepared from organic salt precursors. International Journal of Hydrogen Energy. 2012;**37**:16375-16381

[7] Papadopoulou E, Ioannides T. Methanol Reforming over Cobalt Catalysts Prepared from Fumarate Precursors: TPD Investigation. Catalysts. 2016;**6**:33

[8] Papadopoulou E, Ioannides T. Steam reforming of methanol over cobalt catalysts: Effect of cobalt oxidation state. International Journal of Hydrogen Energy. 2015;**40**:5251-5255

[9] Semenov SA, Drobot DV, Musatova VY, Pronin AS, Pomogailob AD, Dzhardimalieva GI, Popenko VI. Synthesis and thermal conversions of unsaturated cobalt(II)dicarboxylates as precursors of metallopolymer nanocomposites. Russian Journal of Inorganic Chemistry. 2015;**60**:897-905

[10] Mohamed MA, Galwey AK, Halawy SA. Kinetic and thermodynamic study of the non-isothermal decompositions of cobalt malonate dihydrate and of cobalt hydrogen malonate dihydrate. Thermochimica Acta. 2000;**346**:91-103

[11] Dippong T, Berinde Z, Pauliuc I. Formation and evolution of crystalline magnetic phases of $CoFe_2O_4$ with temperature depending on the nature of the diol used. Journal of Magnetism and Magnetic Materials. 2013;**334**:87-95

[12] Manouchehri I, Kameli P, Salamati H. Facile Synthesis of Co_3O_4/CoO Nanoparticles by thermal treatment of ball-milled precursors. Journal of Superconductivity and Novel Magnetism. 2011;**24**:1907-1910

[13] Ganguly A, Kundu R, Ramanujachary KV, Lofland SE, Das D, Vasanthacharya NY, Ahmad T, Ganguli AK. Role of carboxylate ion and metal oxidation state on the morphology and magnetic properties of nanostructured metal carboxylates and their decomposition products. Journal of Chemical Sciences. 2008;**120**:521-528

[14] Abdelkader A, Daly H, Saih Y, Morgan K, Mohamed MA, Halawy SA, Hardacre C. Steam reforming of ethanol over Co_3O_4-Fe_2O_3 mixed oxides. International Journal of Hydrogen Energy. 2013;**38**:8263-8275

[15] de Rivas B, López-Fonseca R, Jiménez-González C, Gutiérrez-Ortiz JI. Highly active behaviour of nanocrystalline Co_3O_4 from oxalate nanorods in the oxidation of chlorinated short chain alkanes. Chemical Engineering Journal, 2012;**184**:184-192

[16] Shen Y, Zhang Z, Xiao K. Evaluation of cobalt oxide, copper oxide and their solid solutions as heterogeneous catalysts for Fenton-degradation of dye pollutants. RSC Advances. 2015;**5**:91846-91854

[17] Manteghi F, Kazemi SH, Peyvandipoor M, Asghari A. Preparation and application of cobalt oxide nanostructures as electrode materials for electrochemical supercapacitors. RSC Advances. 2015;**5**:76458-76463

[18] Yuan W, Xie D, Dong Z, Su Q, Zhang J, Du G, Xu B. Preparation of porous Co_3O_4 polyhedral architectures and its application as anode material in lithium-ion battery. Materials Leters. 2013;**97**:129-132

[19] Yuan Z, Huang F, Feng C, Sun J, Zhou Y. Synthesis and electrochemical performance of nanosized Co_3O_4. Materials Chemistry and Physics. 2003;**79**:1-4

[20] de Rivas B, López-Fonseca R, Jiménez-González C, Gutiérrez-Ortiz JI. Synthesis, characterisation and catalytic performance of nanocrystalline Co_3O_4 for gas-phase chlorinated VOC abatement. Journal of Catalysis. 2011;**281**:88-97

[21] Liu Q, Wang L-C, Chen M, Cao Y, He H-Y, Fan K-N. Dry citrate-precursor synthesized nanocrystalline cobalt oxide as highly active catalyst for total oxidation of propane. Journal of Catalysis. 2009;**263**:104-113

[22] Wang C, Zhang C, Hua W, Guo Y, Lu G, Gil S, Giroir-Fendler A. Catalytic oxidation of vinyl chloride emissions over Co-Ce composite oxide catalysts. Chemical Engineering Journal. 2017;**315**:392-402

[23] Pudukudy M, Yaakob Z, Narayanan B, Gopalakrishnan A, Tasirin SM. Facile synthesis of bimodal mesoporous spinel Co_3O_4 nanomaterials and their structural properties. Superlatices and Microstructures. 2013;**64**:15-26

[24] Stefanescu M, Dippong T, Stoia M, Stefanescu O. Study on the obtaining of cobalt oxides by thermal decomposition of some complex combinations, undispersed and dispersed in SiO_2 matrix. Journal of Thermal Analysis and Calorimetry. 2008;**94**:389-393

[25] Bhatacharjee CR, Purkayastha DD, Das N. Surfactant-free thermal decomposition route to phase pure tricobalt tetraoxide nanoparticles from cobalt(II)-tartrate complex. Journal of Sol-Gel Science and Technology. 2013;**65**:96-300

[26] Palacios-Hernández T, Hirata-Flores GA, Contreras-López OE, Mendoza-Sánchez ME, Valeriano-Arreola I, González-Vergara E, Méndez-Rojas MA. Synthesis of Cu and Co metal oxide nanoparticles from thermal decomposition of tartrate complexes. Inorganica Chimica Acta. 2012;**392**:277-282

[27] Chen Z, Xu A, Yu Zhang NG. Preparation of NiO and CoO nanoparticles using M^{2+}-oleate (M = Ni, Co) as precursor. Current Applied Physics. 2010;**10**:967-970

[28] Assim K, Schulze S, Pügner M, Uhlemann M, Gemming T, Giebeler L, Hietschold M, Lampke T, Lang H. Co(II) ethylene glycol carboxylates for Co_3O_4 nanoparticle and nanocomposite formation. Journal of Materials Science. 2017;**52**:6697-6711

[29] Gawande MB, Pandeyb RK, Jayaram RV. Role of mixed metal oxides in catalysis science – versatile applications in organic synthesis. Catalysis Science & Technology. 2012;**2**:1113-1125

[30] Diodati S, Nodari L, Natile MM, Caneschi A, de Julián Fernández C, Hoffmann C, Kaskel S, Lieb A, Di Noto V, Mascoto S, Saini R, Gross S. Coprecipitation of oxalates: An easy and reproducible wet-chemistry synthesis route for transition-metal ferrites. European Journal of Inorganic Chemistry. 2014;**5**:875-887

[31] Dippong T, Levei EA, Cadar O. Preparation of $CoFe_2O_4/SiO_2$ nanocomposites at low temperatures using short chain diols. Journal of Chemistry, Hindawi. 2017;**2017**:Article ID: 7943164

[32] Stefanescu M, Stoia M, Caizer C, Dippong T, Barvinschi P. Preparation of $Co_xFe_{3-x}O_4$ nanoparticles by thermal decomposition of some organo-metallic precursors. Journal of Thermal Analysis and Calorimetry. 2009;**97**:245-250

[33] Barvinschi P, Stefanescu O, Dippong T, Sorescu S, Stefanescu M. $CoFe_2O_4/SiO_2$ nanocomposites by thermal decomposition of some complex combinations embedded in hybrid silica gels. Journal of Thermal Analysis and Calorimetry. 2013;**112**:447-453

[34] Dippong T, Levei EA, Cadar O, Mesaros A, Borodi G. Sol-gel synthesis of $CoFe_2O_4$:SiO_2 nanocomposites – insights into the thermal decomposition process of precursors. Journal of Analytical and Applied Pyrolysis. 2017;**125**:169-177

[35] Dippong T, Levei EA, Cadar O, Goga F, Borodi G, Barbu-Tudoran L. Thermal behavior of $Co_xFe_{3-x}O_4/SiO_2$ nanocomposites obtained by a modified sol-gel method. Journal of Thermal Analysis and Calorimetry. 2017;**128**:39-52

[36] Amiri M, Niasari MS, Akbari A, Razavi R. Sol-gel auto-combustion synthesize and characterization of a novel anticorrosive cobalt ferrite nanoparticles dispersed in silica matrix. Journal of Materials Science: Materials in Electronics. 2017;**28**:10495-10508

[37] Muntean C, Bozdog M, Duma S, Stefanescu M. Study on the formation of $Co_{1-x}Zn_xFe_2O_4$ system using two low-temperature synthesis methods. Journal of Thermal Analysis and Calorimetry. 2016;**123**:117-126

[38] Stefanescu M, Bozdog M, Muntean C, Stefanescu O, Vlase T. Synthesis and magnetic properties of $Co_{1-x}Zn_xFe_2O_4$ (x = 0-1) nanopowders by thermal decomposition of Co(II), Zn(II) and Fe(III) carboxylates. Journal of Magnetism and Magnetic Materials. 2015;**393**:92-98

[39] Faure B, Alphonse P. Co–Mn-oxide spinel catalysts for CO and propane oxidation at mild temperature. Applied Catalysis. B, Environmental. 2016;**180**:715-725

[40] Wang C, Liu S, Liu L, Bai X. Synthesis of cobalt-aluminate spinels via glycine chelated precursors. Materials Chemistry and Physics. 2006;**96**:361-370

[41] Gingasu D, Mindru I, Culita DC, Patron L, Calderon-Moreno JM, Osiceanu P, Preda S, Oprea O, Parvulescu V, Teodorescu V, Walsh JPS. Structural, magnetic and catalytic properties of cobalt chromite obtained through precursor method. Materials Research Bulletin. 2015;**62**:52-64

[42] Darbar D, Reddy MV, Sundarrajan S, Patabiraman R, Ramakrishna S, Chowdari BVR. Anodic electrochemical performances of $MgCo_2O_4$ synthesized by oxalate decomposition method and electrospinning technique for Li-ion battery application. Materials Research Bulletin. 2016;**73**:369-376

[43] Kim BC, Rajesh M, Jang HS, Yu KH, Kim SJ, Park SY, Raj CJ. Facile synthesis and capacitive properties of nickel-cobalt binary metal oxide nanoaggregates via oxalate route. Journal of Alloys and Compounds. 2016;**674**:376-383

[44] Seyfi B, Baghalha M, Kazemian H. Modified $LaCoO_3$ nano- perovskite catalysts for the environmental application of automotive CO oxidation. Chemical Engineering Journal. 2009;**148**:306-311

[45] Gabal MA, Hameed SA, Obaid AY. $CoTiO_3$ via cobalt oxalate–TiO_2 precursor. Synthesis and characterization. Materials Characterization. 2012;**71**:87-94

[46] Zhou H, Kitagawa S. Metal-Organic Frameworks (MOFs). Chemical Society Reviews. 2014;**43**:5415

[47] Guo H, Li X, Li XH, Wang Z-X, Peng W, Sun Q, Xie J. Preparation and electrochemical properties of $Co3O_4$/graphite composites as anodes of lithium ion batteries. Journal of Central South University of Technology. 2010;**17**:498-503

[48] Wang W, Li Y, Zhang R, He D, Liu H, Liao S. Metal-organic framework as a host for synthesis of nanoscale $Co3O_4$ as an active catalyst for CO oxidation. Catalysis Communications. 2011;**12**:875-879

[49] Ashouri F, Zare M, Bagherzadeh M. Manganese and cobalt-terephthalate metal-organic frameworks as a precursor for synthesis of Mn_2O_3, Mn_3O_4 and Co_3O_4 nanoparticles: Active catalysts for olefin heterogeneous oxidation. Inorganic Chemistry Communications. 2015;**61**:73-76

[50] Pei Y, Li Z, Li Y. Highly Active and Selective Co-Based Fischer-Tropsch Catalysts Derived from Metal–Organic Frameworks. AICHE Journal. 2017;**63**(7):2935-2944

[51] Zhou P, Zhang Z, Jiang L, Yu C, Lv K, Sun J, Wang S. A versatile cobalt catalyst for the reductive amination of carbonyl compounds with nitro compounds by transfer hydrogenation. Applied Catalysis. B, Environmental. 2017;**210**:522-5329

[52] Khan IA, Nasim F, Choucair M, Ullah S, Badshah A, Nadeem MA. Cobalt oxide nanoparticles embedded N-CNTs: Lithium ion battery applications. RSC Advances. 2016;**6**(2):1129-1135

[53] Lu H, Zhang H, Liu R, Zhang X, Zhao H, Wang G. Macroscale cobalt-MOFs derived metallic Co nanoparticles embedded in N-doped porous carbon layers as efficient oxygen electrocatalysts. Applied Surface Science. 2017;**392**:402-409

[54] Qiu B, Guo W, Liang Z, Xia W, Gao S, Wang Q, Yu X, Zhao R, Zou R. Fabrication of Co_3O_4 nanoparticles in thin porous carbon shells from metal-organic frameworks for enhanced electrochemical performance. RSC Advances. 2017;**7**:13340-13346

[55] Huang Q, Zhang J, He Z, Shi P, Qin X, Yao W. Direct fabrication of lamellar self-supporting Co_3O_4/N/C peroxymonosulfate activation catalysts for effective aniline degradation. Chemical Engineering Journal. 2017;**313**:1088-1098

[56] Dippong T, Levei EA, Borodi G, Goga F, Tudoran LB. Influence of Co/Fe ratio on the oxide phases in nanoparticles of $Co_xFe_{3-x}O_4$. Journal of Thermal Analysis and Calorimetry. 2015;**119**:1001-1009

Chapter 4

Hydrothermally Produced Cobalt Oxide Nanostructures at Different Temperatures and Effect on Phase Transition Temperature and Threshold Voltage of Nematic Liquid Crystal Host

Hasan Eskalen, Süleyman Kerli and Şükrü Özgan

Additional information is available at the end of the chapter

http://dx.doi.org/10.5772/intechopen.70946

Abstract

In this study, cobalt oxide Co_3O_4 nanostructured material is synthesized by hydrothermal method by using different concentration of cobalt acetate salt at unique hydrothermal reaction time for five different hydrothermal reaction temperatures 105, 120, 140, 160, and 180°C. The obtained nanoparticles are annealed at 300°C for 5 h. Co_3O_4 nanostructures are determined by means of scanning electron microscopy (SEM), X-ray diffraction (XRD), and UV-vis spectroscopy. The hydrothermally produced samples were reannealed at 550°C, and morphological and structural properties were investigated deeply again. The effect of annealing temperature on morphologies and crystalline structure of cobalt oxide nanoparticle (NP) was also investigated. Nanopyramids and nanorods are two main morphologically obtained structures from the hydrothermal experiment. Nematic liquid crystal mixture E7 is doped with Co_3O_4 nanorod. Phase transition and threshold voltage of pure and Co_3O_4 NP-doped E7 LC are examined successfully. It reveals that for Co_3O_4 NP-doped E7 phase transition temperature and threshold voltage increased very slightly.

Keywords: cobalt oxide, hydrothermal synthesis, band gap, threshold voltage, thermal property

1. Introduction

"There is Plenty of Room at the Bottom" a titled lecture in 1959 was given by Richard Feynman, and this has opened a new era in the field of nanotechnology [1, 2], which can be defined as the engineering of functional systems at the molecular level [3]. Nanoscience and nanotechnology have been growing rapidly due to the fact that nanomaterials are synthesized with new strategies and those synthesized materials are characterized and manipulated with new tools [4]. Nanostructures can be classified into three different categories according to their size and shapes: zero dimensional (0D), one dimensional (1D), and two dimensional (2D) [5, 6]. Nanodots-nanospheres, nanowires-nanorods, and nanosheets-nanoplates are some examples of different shapes and size nanomaterials for this classification. Nanomaterials, unlike to conventional materials, can be physically and chemically manipulated and used in different areas [7, 8] including biology and medicine [9–11], water treatment [12], electronics [13, 14], and optics [15–17].

Magnetic fields more or less impact a certain subclass of NP, which is called as magnetic nanoparticles (MNPs) [18, 19]. MNPs have their own special properties, such as superparamagnetism [20], high mass transference [21], and high field irreversibility [22]. The second most popular MNP, considering to application areas, is cobalt oxide (Co_3O_4) since these nanoparticles are used in various fields from micro-electronics to drug delivery [7]. Co_3O_4 is one of the significant transition metal oxides, and the direct optical band gap of bulk Co_3O_4 is 2.19 eV.

Dispersion of nanomaterials into liquid crystals (LCs) has been a topic of great interest in recent years. There are many distinguished researchers investigating liquid crystals, and some of whose works are focused on doping nanomaterials into liquid crystals [23–29]. Gold, silver, zinc oxide, carbon, and quantum dots nanoparticles are some of the examples of guest nanoparticles. Co_3O_4 nanomaterials doped nematic liquid crystals (NLC) have only been investigated for spherical morphological Co_3O_4 nanoparticles to our knowledge [30].

Researchers have focused on synthesizing Co_3O_4 nanoparticles not only with diverse morphology but also with different methods since any change in production methods, particle size, shape, and structure of Co_3O_4 nanomaterials could lead producing new nanomaterials for potential applications with unique properties. Up to now, Co_3O_4 nanoparticles with various morphologies such as nanospheres [31, 32], nanorods [33–36], nanowalls [37], nanoneedles [38], nanobelts [39], nanowires [40], nanoflowers [41], nanotubes [42, 43], nanofibers [44], nanodiscs [45], and nanochains [46] have been synthesized with different approaches, including urea precipitation [47], chemical vapor deposition [48], sol-gel [49], microwave-assisted process [50], wet chemical approach [51], and hydrothermal method [39, 52–54]. Hydrothermal methods are widely used to produce various different Co_3O_4 nanomaterials. Easily scaling up to industrial demand, requiring low temperature, no need for calcination, being inexpensive, and having a fairly uniform particle size and morphology are some of the advantages of this nanoparticle production method [55]. The term hydrothermal is used when water is used as a solvent, and solvothermal is used when organics are used as solvent [56]. In literature, the effect of hydrothermal reaction times and reaction temperature on morphologies of Co_3O_4 nanoparticles was examined [57–59].

In this study, new morphologies of Co_3O_4 have been synthesized by hydrothermal method using Cobalt(II) chloride hexahydrate precursor at different unique temperatures and hydrothermal reaction times. The influence of hydrothermal temperatures on crystal structures and particle morphologies was examined deeply. Five different hydrothermal temperature points were selected between 105 and 180°C, and the effect of annealing temperatures was also investigated. The obtained particles were firstly annealed at 300°C for 5 h and then reannealed 500°C. The prepared Co_3O_4 NPs were characterized by XRD, SEM, and UV-vis. Moreover, we selected Co_3O_4 NPs obtained at 120°C hydrothermal reaction temperature to investigate how nanorod morphological Co_3O_4 NPs affect nematic liquid crystal phase transition and threshold voltages.

2. Materials and methods

2.1. Synthesis of Co_3O_4 NPs

All chemicals were used without further purification. A schematic diagram of the synthesis step is given in **Figure 1**. For the synthesis of the samples, 4.3983 g Cobalt(II) chloride hexahydrate

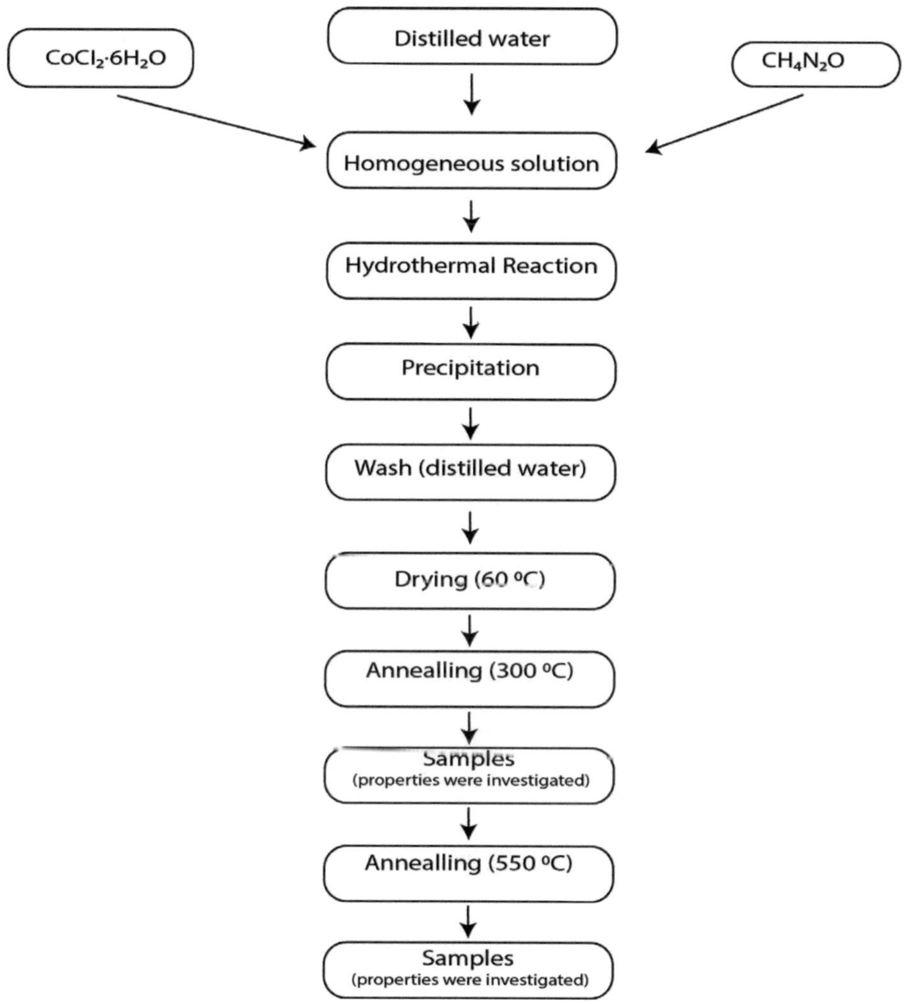

Figure 1. Schematic diagram for the synthesis procedure of Co_3O_4.

and 0.12 g Urea (CH_4N_2O) solution were mixed with 50 ml distilled water. The prepared solution was placed in the autoclave, and it waited in the furnace at 105, 120, 140, 160, and 180°C for 6 h. It was precipitated and washed several times with distilled water and dried at 60°C for 10 h. The obtained particles were annealed for 5 h at 300°C. Moreover, all obtained samples were also annealed at 500°C for 1 h to compare the effects of different annealing temperatures on XRD and SEM results of obtained Co_3O_4 samples.

2.2. Liquid crystal experiment

The LC used was commercially available eutectic mixture E7 (SYNTHON Chemicals, Germany). The nematic to isotropic transition temperature for pure E7 is measured by using polarizing optical microscope integrated with hot stage, and it is found 58.6°C. The liquid crystal cells were made by indium-tin oxide-coated optical glass plates with a planar alignment layers. The thicknesses of cells were about 8 μm and an effective area of 1 cm2. The cells were purchased from Instec, Inc. Hydrothermally produced Co_3O_4 sample at 120°C was selected to mixed in host E7 since the morphology of this sample contains nanorods. E7 nematic LC mixture and Co_3O_4 nanoparticles were dissolved in isopropanol followed by the ultrasonic bath. The mixture was left for 48 h in the furnace at 50°C to fully evaporate isopropanol. 0.05% doped sample was filled to LC cell at 60°C by capillarity action [25].

2.3. Characterization

The hydrothermally obtained Co_3O_4 samples' crystalline structures were investigated by a Philips X' Pert Pro X-ray diffractometer (XRD) with Cu-Kα radiation. The morphologies of the samples were examined by scanning electron microscope (Zeiss EVO 10LS). The optical

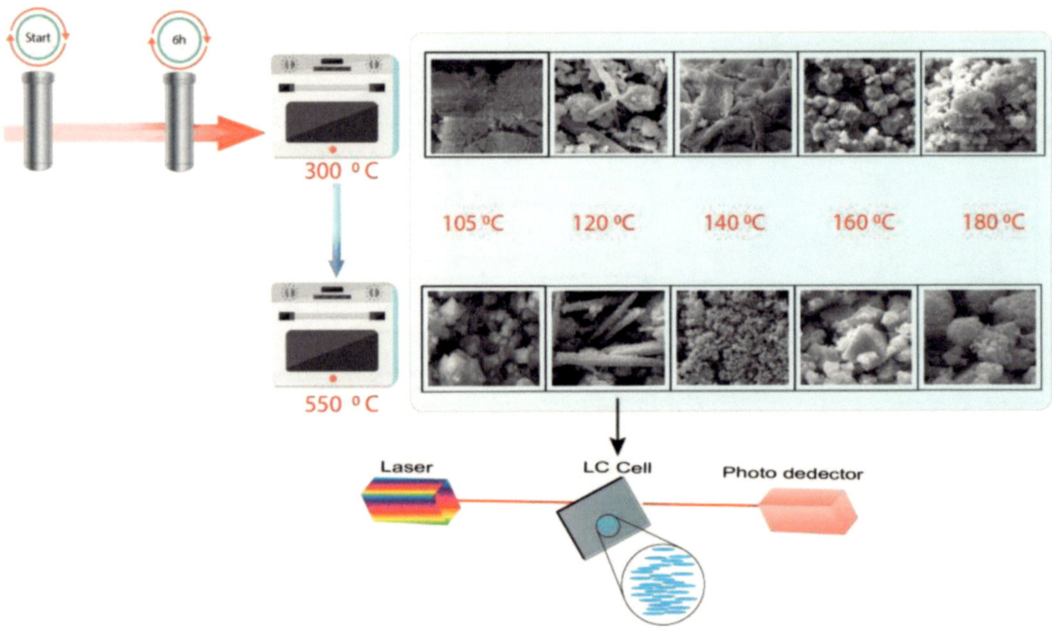

Figure 2. Synthesis of Co_3O_4 samples and basic setup of LC light transmittance experiment.

http://dx.doi.org/10.5772/intechopen.70946

property samples were obtained by a Shimadzu UV-1800 ultraviolet visible spectroscopy (UV-vis) in the range of 200–900 nm using distilled water as a reference solvent. The samples were dispersed into distilled water solvent and ultrasonicated before the measurement. The textures of E7 and E7-doped sample were taken using polarizing optical microscope (Eclipse E200, Nikon, Japan) equipped with the digital camera. Temperature was controlled with heating stage (LTS 120, LinkamScientific Instruments, Ltd., England) with a temperature accuracy of ± 0.1°C controlled with PE95 LinkPad. Optical transmittance experiment designed and performed by laser, polarizer, analyzer, and photodiode and experimental scheme is described in detail in the literature [60]. Experimental detail about synthesis of Co_3O_4 samples and basic setup of liquid crystal optical transmittance experimental is given in **Figure 2**.

3. Results and discussion

X-ray diffraction pattern of Co_3O_4 samples annealed at 300°C for 6 h is given in **Figure 3**. The diffraction peaks of hydrothermally produced cubic structured sample at 120°C are suitable with the values in the standard card (PDF-2, reference code: 01-074-1656) and the Co_3O_4 particles produced at 180°C are partly suitable to this reference code as shown in **Figure 3**. The others produced Co_3O_4 particles' apparent crystal structure patents were not observed. It is supposed that this was originated from hydroxide structures and chemical contaminant.

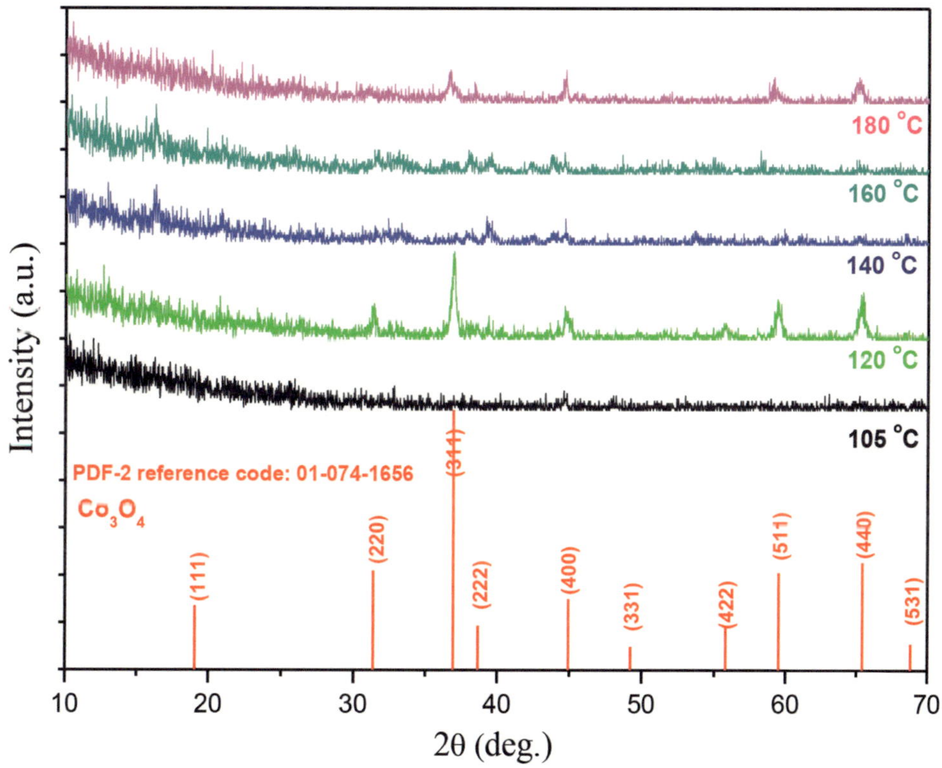

Figure 3. XRD graph of Co_3O_4 samples annealed at 300°C.

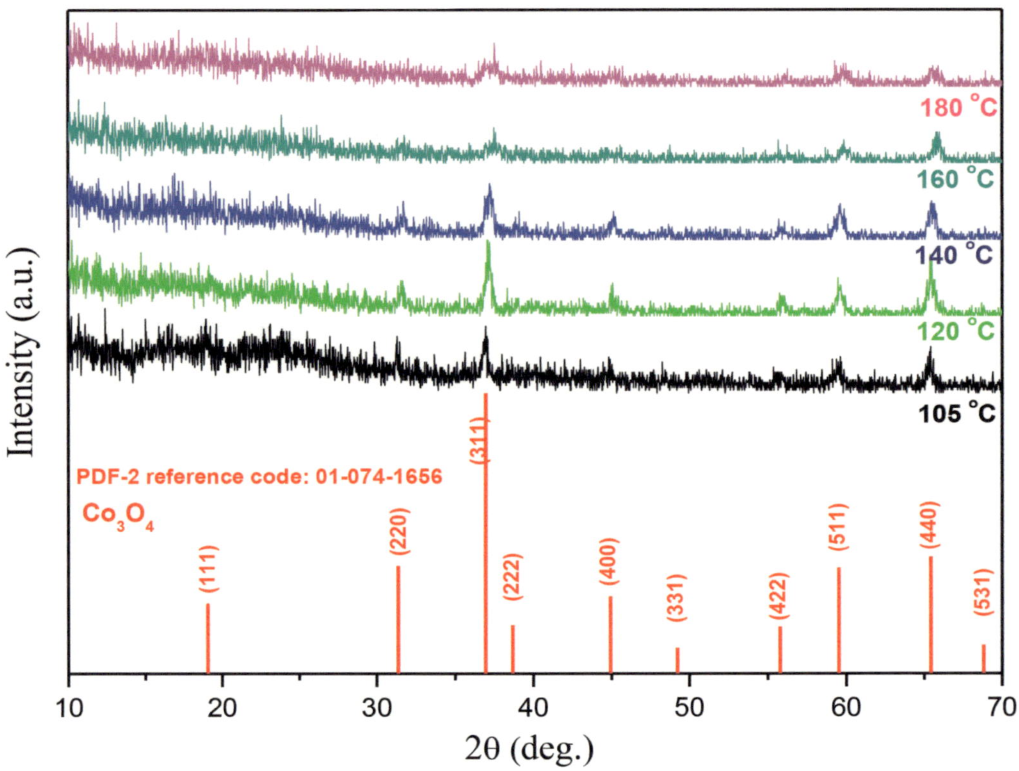

Figure 4. XRD graph of Co_3O_4 samples annealed at 550°C.

The obtained Co_3O_4 particles reannealed at 550°C for an hour and crystal structure of reannealed samples were investigated again. The effect of annealing temperature on the crystalline structure of obtained particles is shown in **Figure 4**. In this figure, XRD peaks of Co_3O_4 particles become clear, and all samples give at least some of the reference peaks of Co_3O_4. Hydroxide structures disappeared with reannealing at 550°C, and XRD peaks of Co_3O_4 particles are seen more clearly than samples annealed at 300°C.

The morphologies of particles that are produced using hydrothermal method at various temperatures and annealed at 300°C for 5 h are seen in **Figure 5**. In this figure, hydroxide and chemical waste structures are observed. Especially in **Figure 5b**, nanorod structures are seen. Nanosheet structures as layers are seen in **Figure 5d**. However, remarkable structures have not been observed for the samples produced at 105, 140, and 180°C, which are corresponding to **Figure 5a**, **c**, and **e**.

To observe the effect of annealing temperature, SEM pictures belonging to particles annealed within 550°C for 1 h are given in **Figure 6**. It was observed that the hydroxide compounds and chemical wastes decreased with the effect of the annealing. In **Figure 6a** and **d**, octahedral microparticles have been observed clearly, and in **Figure 6b**, nano−/microrods have been shown. **Figure 6c** shows the octahedral particles in nano level. In **Figure 6e**, agglomerated nanoparticles are dominantly seen.

Figure 5. Morphologies of 300°C annealed samples produced at (a) 105°C, (b) 120°C, (c) 140°C, (d) 160°C, and (e) 180°C.

Figure 6. Morphologies of 550°C annealed samples produced at (a) 105°C, (b) 120°C, (c) 140°C, (d) 160°C, and (e) 180°C.

The optical bad gap (Eg) for the direct transition was obtained using Tauc plot. Optical absorption graph of sampled annealed at 300°C is given in **Figure 7**, left column and optical absorption of sampled reannealed at 550°C is given in **Figure 7**, right column. The direct band gap of 300°C annealed Co_3O_4 NP changing between 3.1 and 3.5; on the other hand, reannealed sampled band gaps were very close to 3.5 eV as interpreted in **Figure 7**. The measurement results of direct band gaps of reannealed nanoparticles were suitable with the literature [61, 62].

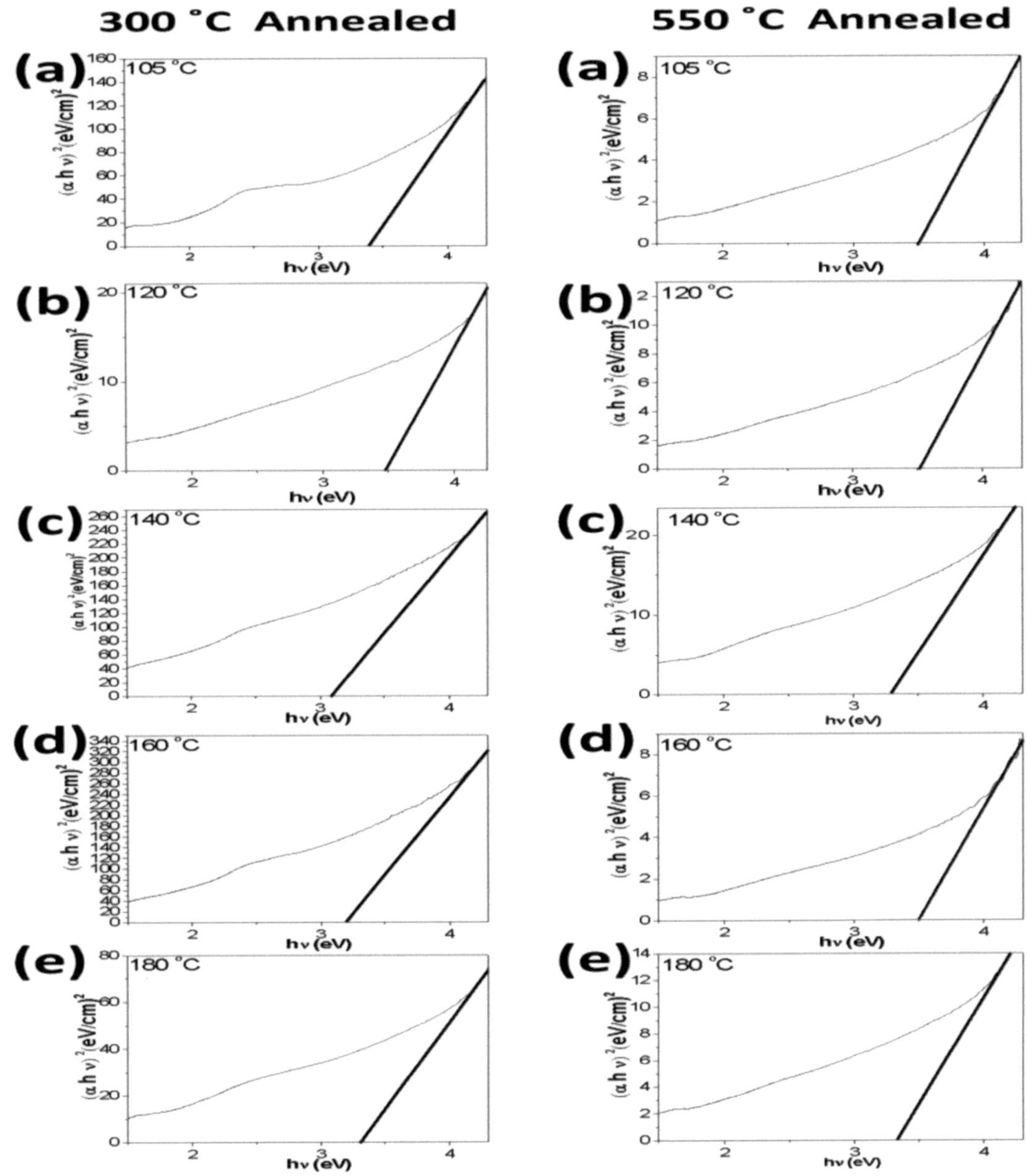

Figure 7. UV-vis spectra and band gaps of hydrothermally obtained particles annealed at 300 and 550°C.

Polarizing optical microscope was used to investigate texture of pure and 0.05% nanoparticle-doped E7. The different magnification textures of pure E7 are given in **Figure 8**. The smallest magnification ratio of pure E7 is given in **Figure 8a**, and the most detailed texture of this LC is shown in **Figure 8d**.

Phase transitions of 0.05% cobalt oxide nanoparticle-doped E7 LC are illustrated in **Figure 9a**. The first drop of isotropic liquid appeared at 57.1°C, T_N, and the last drop nematic disappeared

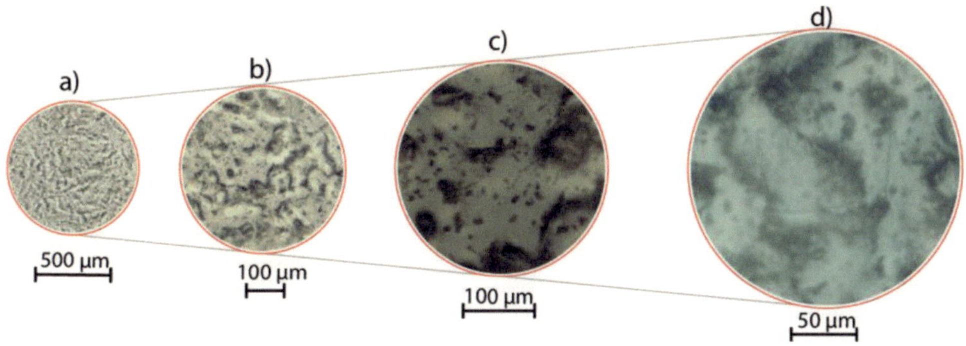

Figure 8. Optical polarizing image of E7 nematic LC, (a) 40× magnification, (b) 100× magnification, (c) 200× magnification, and (d) 400× magnification.

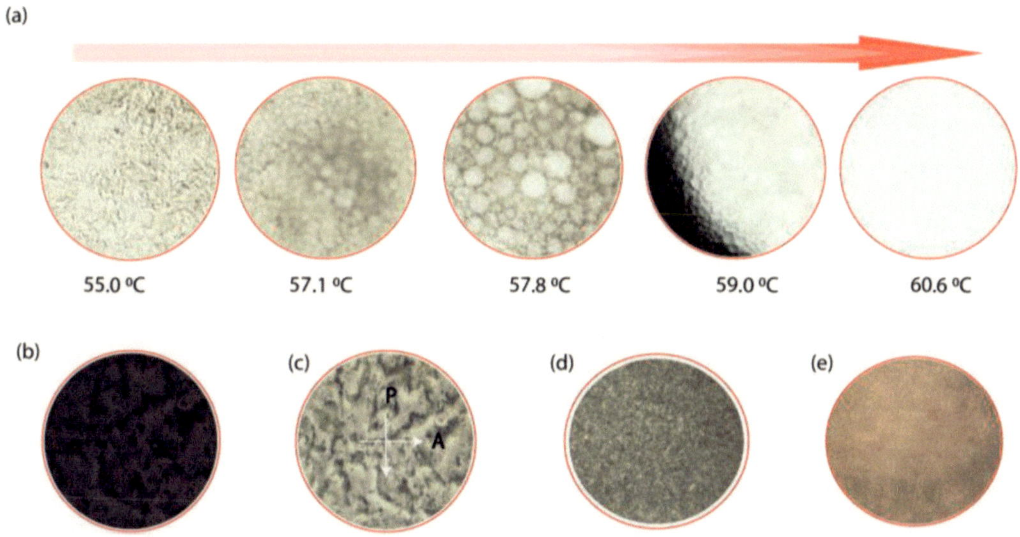

Figure 9. Texture of pure and NP-doped E7, (a) phase transition, (b) without cross polarizer, (c) with cross polarizer, and (d) E7 at 1 V, (e) 0.05% NP-doped E7 at 1 V.

at 60.6°C, T_I. The average temperature of nematic-isotropic transition of doped sample is calculated by using equation $T_{NI} = 0.5(T_N + T_I)$ [63], and found 58.85°C. The phase transition difference of pure E7 and NP-doped E7 is found as $\Delta T_{NI} = 0.2$°C. **Figure 9b** and **c** shows the optical picture of E7 LC without and with cross polarizers, respectively. The texture of pure and doped LC under 1 V applied voltage is given in **Figure 9d** and **e**.

Transmission versus voltage graph of pure and Co_3O_4 NP-doped LC is illustrated in **Figure 10**. The wavelength of used laser light in transmission experiment was 650 nm. The prepared LC cells are placed between cross polarizers, and the angle between cross polarizers and LC cell is adjusted to 45°, and the output signal was detected by a photodiode detector. Transmission voltage behavior of pure and doped sample is not very different from each other, which

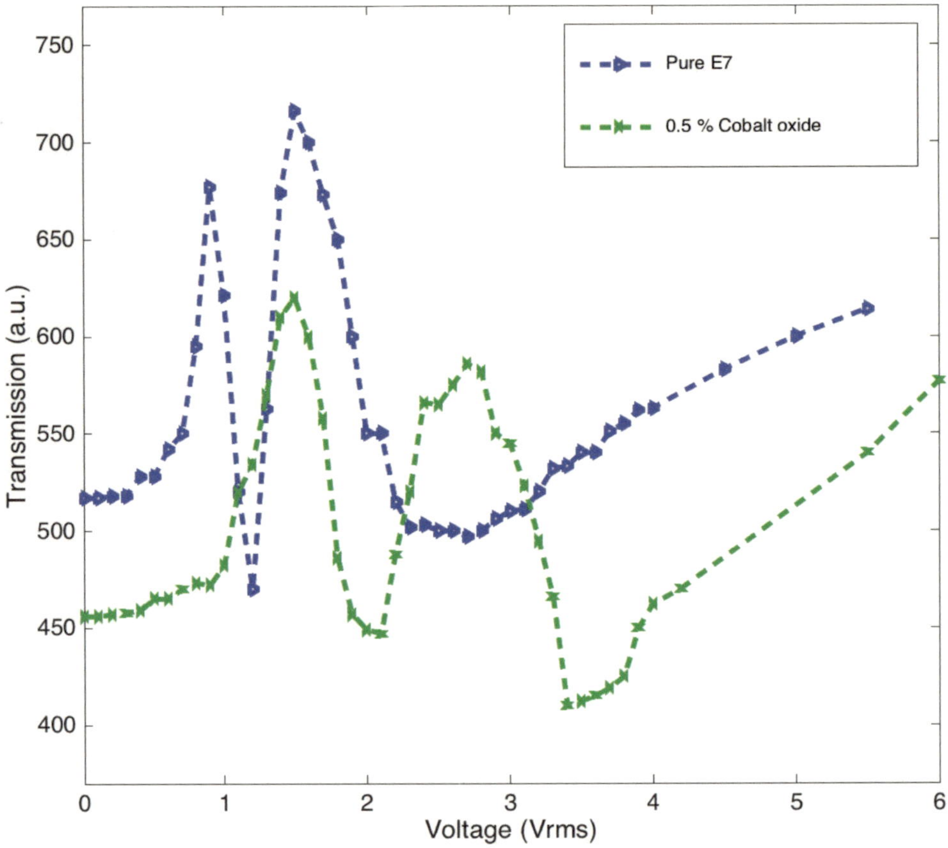

Figure 10. Transmission vs. voltage graph for pure and NP-doped E7.

implies that threshold voltage values are close to each other. The calculated threshold voltage for pure E7 and 0.05% Co_3O_4 NP-doped E7 are 0.8 V and 0.9 V sequentially.

4. Conclusion

In summary, Co_3O_4 nanostructures were synthesized using the hydrothermal method with unique hydrothermal reaction time at different temperatures. The crystalline structures, morphologies, and optical absorptions were investigated in detail with XRD, SEM, and UV-vis spectroscopy for two different annealed temperatures. The obtained samples were firstly investigated after being annealed at 300°C for 5 h, and the results of SEM and XRD separately indicated that cobalt hydroxide did not decompose fully to form cobalt oxide nanocrystalline. The samples were reannealed at 500°C for an hour to investigate deeply. The hydroxide compounds and chemical wastes are removed, and XRD peaks of Co_3O_4 particles become clear. Obtained morphologies of cobalt oxide nanostructures also changed with calcination. The particles, with a nanorod morphology, produced at 120°C hydrothermal reaction temperature and annealed at 500°C were selected to doped nematic LC mixture E7. Phase transition and the threshold voltage of pure and Co_3O_4 NP-doped E7 LC were examined successfully. It reveals that for Co_3O_4 NP-doped E7 phase transition temperature and threshold voltage increased very slightly.

Acknowledgements

The authors thank Kahramanmaras Sutcu Imam University for the continuous motivation to maintain this research. This work was financially supported by Kahramanmaras Sutcu Imam University, (KSU) Scientific Research Projects Coordination Department, under Project No. 2017/4-29D.

Author details

Hasan Eskalen[1]*, Süleyman Kerli[2] and Şükrü Özgan[3]

*Address all correspondence to: heskalen@gmail.com

1 Material Science and Engineering, Graduate School of Natural and Applied Science, Kahramanmaras Sutcu Imam University, Kahramanmaras, Turkey

2 Energy System Engineering Department, Kahramanmaras Sutcu Imam University, Kahramanmaras, Turkey

3 Department of Physics, Faculty of Science and Letter, Kahramanmaras Sutcu Imam University, Kahramanmaras, Turkey

References

[1] Feynman RP. There's plenty of room at the bottom. Engineering and Science. 1960;**23**(5): 22-36

[2] Sanjay SS, Pandey AC. A brief manifestation of nanotechnology. In: EMR/ESR/EPR Spectroscopy for Characterization of Nanomaterials. Springer, India. 2017. p. 47-63

[3] Karkare M. Nanotechnology: Fundamentals and Applications. IK International Pvt Ltd; 2008

[4] Rao CNR, Müller A, Cheetham AK. The Chemistry of Nanomaterials: Synthesis, Properties and Applications. Germany: John Wiley & Son; 2006

[5] Nakato T, Kawamata J, Takagi S. Inorganic Nanosheets and Nanosheet-Based Materials: Fundamentals and Applications of Two-Dimensional Systems. Springer, Japon. 2017

[6] Yu Z et al. Supercapacitor electrode materials: nanostructures from 0 to 3 dimensions. Energy & Environmental Science. 2015;**8**(3):702-730

[7] Sundar LS, Anjum NA, Ferro MC, Pereira E, Singh MK, Sousa ACM. (2017). Biocompatibility and biotoxicity of in-situ synthesized carboxylated nanodiamond-cobalt oxide nanocomposite. Journal of Materials Science & Technology.

[8] Oberdörster G, Oberdörster E, Oberdörster J. Nanotoxicology: An emerging discipline evolving from studies of ultrafine particles. Environmental Health Perspectives. 2005;**113**(7):823-839

[9] Salata OV. Applications of nanoparticles in biology and medicine. Journal of Nanobiotechnology. 2004;**2**(1):3

[10] Wang AZ, Langer R, Farokhzad OC. Nanoparticle delivery of cancer drugs. Annual Review of Medicine. 2012;**63**:185-198

[11] Peer D et al. Nanocarriers as an emerging platform for cancer therapy. Nature Nanotechnology. 2007;**2**(12):751-760

[12] Hu J, Chen G, Lo IM. Removal and recovery of Cr (VI) from wastewater by maghemite nanoparticles. Water Research. 2005;**39**(18):4528-4536

[13] Maier SA et al. Local detection of electromagnetic energy transport below the diffraction limit in metal nanoparticle plasmon waveguides. Nature Materials. 2003;**2**(4):229-232

[14] Wong WS, Salleo A. Flexible Electronics: Materials and Applications. Vol. 11. Springer US: Science & Business Media; 2009

[15] Barnes WL, Dereux A, Ebbesen TW. Surface plasmon subwavelength optics. Nature. 2003;**424**(6950):824-830

[16] Gavrilenko VI. Optics of Nanomaterials. Pan Stanford Publishing. Singapore; 2011

[17] Martin, Charles R. Nanomaterials: a membrane-based synthetic approach. Science (1994): 1961-1966

[18] Gubin SP. Magnetic Nanoparticles. Germany: John Wiley & Sons; 2009

[19] Issa B et al. Magnetic nanoparticles: Surface effects and properties related to biomedicine applications. International Journal of Molecular Sciences. 2013;**14**(11):21266-21305

[20] Tartaj P et al. The preparation of magnetic nanoparticles for applications in biomedicine. Journal of Physics D – Applied Physics. 2003;**36**(13):R182-R197

[21] Rocha-Santos TA. Sensors and biosensors based on magnetic nanoparticles. TrAC Trends in Analytical Chemistry. 2014;**62**:28-36

[22] Kodama R. Magnetic nanoparticles. Journal of Magnetism and Magnetic Materials. 1999; **200**(1):359-372

[23] Hegmann T, Qi H, Marx VM. Nanoparticles in liquid crystals: Synthesis, self-assembly, defect formation and potential applications. Journal of Inorganic and Organometallic Polymers and Materials. 2007;**17**(3):483-508

[24] Eskalen H et al. Electro-optical properties of liquid crystals composite with zinc oxide nanoparticles. Acta Physica Polonica, A. 2015;**127**(3):756-760

[25] Shiju E et al. Effect of ferroelectric nanoparticles in the alignment layer of twisted nematic liquid crystal display. Optical Materials. 2017;**67**:7-13

[26] Ye L et al. The influence of Ag nanoparticles on random laser from dye-doped nematic liquid crystals. Laser Physics Letters. 2016;**13**(10):105001

[27] Singh U et al. Influence of low concentration silver nanoparticles on the electrical and electro-optical parameters of nematic liquid crystals. Liquid Crystals. 2013;**40**(6):774-782

[28] Zakerhamidi M, Shoarinejad S, Mohammadpour S. Fe_3O_4 nanoparticle effect on dielectric and ordering behavior of nematic liquid crystal host. Journal of Molecular Liquids. 2014;**191**:16-19

[29] Khushboo K et al. Textural, thermal, optical and electrical properties of Iron nanoparticles dispersed 4′-(Hexyloxy)-4-biphenylcarbonitrile liquid crystal mixture. Liquid Crystals. 2017;**44**(11):1717-1726

[30] Chung H-K et al. Superior electro-optic properties of liquid crystal system using cobalt oxide nanoparticle dispersion. Liquid Crystals. 2013;**40**(5):632-638

[31] He T et al. Surfactant-assisted solvothermal synthesis of Co_3O_4 hollow spheres with oriented-aggregation nanostructures and tunable particle size. Langmuir. 2004;**20**(19): 8404-8408

[32] Park J, Shen X, Wang G. Solvothermal synthesis and gas-sensing performance of Co_3O_4 hollow nanospheres. Sensors and Actuators B: Chemical. 2009;**136**(2):494-498

[33] Lian SY et al. Fabrication of single-crystalline Co_3O_4 nanorods via a low-temperature solvothermal process. Materials Letters. 2007;**61**(18):3893-3896

[34] Sun SJ et al. Facile and eco-friendly synthesis of finger-like Co_3O_4 nanorods for electrochemical energy storage. Nanomaterials. 2015;**5**(4):2335-2347

[35] Zhang H et al. From cobalt nitrate carbonate hydroxide hydrate nanowires to porous Co_3O_4 nanorods for high performance lithium-ion battery electrodes. Nanotechnology. 2007;**19**(3):035711

[36] Durano MM, Tamboli AH, Kim H. Cobalt oxide synthesized using urea precipitation method as catalyst for the hydrolysis of sodium borohydride. Colloids and Surfaces A – Physicochemical and Engineering Aspects. 2017;**520**:355-360

[37] Yu T et al. Controlled growth and field-emission properties of cobalt oxide nanowalls. Advanced Materials. 2005;**17**(13):1595-1599

[38] Lou XW et al. Thermal formation of mesoporous single-crystal Co_3O_4 nano-needles and their lithium storage properties. Journal of Materials Chemistry. 2008;**18**(37):4397-4401

[39] Huang H et al. Nanocrystal-constructed mesoporous single-crystalline Co_3O_4 nanobelts with superior rate capability for advanced lithium-ion batteries. ACS Applied Materials & Interfaces. 2012;**4**(11):5974-5980

[40] Gandha K et al. High energy product developed from cobalt nanowires. Scientific Reports. 2014;**4**:1-4

[41] Zhu Z et al. Application of cobalt oxide nanoflower for direct electrochemistry and electrocatalysis of hemoglobin with ionic liquid as enhancer. The Journal of Physical Chemistry C. 2011;**115**(25):12547-12553

[42] Lou XW et al. Self-supported formation of needlelike Co_3O_4 nanotubes and their application as lithium-ion battery electrodes. Advanced Materials. 2008;**20**(2):258-262

[43] Xu J et al. Preparation and electrochemical capacitance of cobalt oxide (Co_3O_4) nanotubes as supercapacitor material. Electrochimica Acta. 2010;**56**(2):732-736

[44] Mao Z et al. Electrospinning synthesis of Co_3O_4@ C nanofibers as a high-performance anode for sodium ion batteries. RSC Advances. 2017;**7**(37):23122-23126

[45] Yang J et al. Synthesis and characterization of cobalt hydroxide, cobalt oxyhydroxide, and cobalt oxide nanodiscs. The Journal of Physical Chemistry C. 2009;**114**(1):111-119

[46] Menezes PW et al. High-performance oxygen redox catalysis with multifunctional cobalt oxide nanochains: Morphology-dependent activity. ACS Catalysis. 2015;**5**(4):2017-2027

[47] Durano MM, Tamboli AH, Kim H. Cobalt oxide synthesized using urea precipitation method as catalyst for the hydrolysis of sodium borohydride. Colloids and Surfaces A: Physicochemical and Engineering Aspects. 2017;**520**:355-360

[48] Koumoulos EP et al. Tribological characterization of chemical vapor deposited Co and Co_3O_4 thin films for sensing reliability in engineering applications. Tribology International. 2015;**82**:89-94

[49] Santos GA et al. Sol-gel synthesis of silica-cobalt composites by employing Co_3O_4 colloidal dispersions. Colloids and Surfaces A: Physicochemical and Engineering Aspects. 2012;**395**:217-224

[50] Jhung SH et al. Microwave effect in the fast synthesis of microporous materials: Which stage between nucleation and crystal growth is accelerated by microwave irradiation? Chemistry-A European Journal. 2007;**13**(16):4410-4417

[51] Li Y, Tan B, Wu Y. Mesoporous Co_3O_4 nanowire arrays for lithium ion batteries with high capacity and rate capability. Nano Letters. 2008;**8**(1):265-270

[52] Liu F et al. Facile synthesis of ultrafine cobalt oxide nanoparticles for high-performance supercapacitors. Journal of Colloid and Interface Science. 2017

[53] Wang G et al. Hydrothermal synthesis and optical, magnetic, and supercapacitance properties of nanoporous cobalt oxide nanorods. The Journal of Physical Chemistry C. 2009;**113**(11):4357-4361

[54] Meher SK, Rao GR. Ultralayered Co_3O_4 for high-performance supercapacitor applications. The Journal of Physical Chemistry C. 2011;**115**(31):15646-15654

[55] Shandilya M, Rai R, Singh J. Review: Hydrothermal technology for smart materials. Advances in Applied Ceramics. 2016;**115**(6):354-376

[56] Li J, Wu Q, Wu J. Synthesis of Nanoparticles via Solvothermal and Hydrothermal Methods. Oak Ridge, TN (United States): Oak Ridge National Laboratory (ORNL); 2015

[57] Yang Y-G et al. Synthesis and characteristics of hierarchical Co_3O_4 powders. Synthesis and Reactivity in Inorganic, Metal-Organic, and Nano-Metal Chemistry. 2016;**46**(3):347-350

[58] Yuanchun Q, Yanbao Z, Zhishen W. Preparation of cobalt oxide nanoparticles and cobalt powders by solvothermal process and their characterization. Materials Chemistry and Physics. 2008;**110**(2):457-462

[59] Ding Y et al. Synthesis and electrochemical properties of Co_3O_4 nanofibers as anode materials for lithium-ion batteries. Materials Letters. 2008;**62**(19):3410-3412

[60] Nayek P, Li G. Superior electro-optic response in multiferroic bismuth ferrite nanoparticle doped nematic liquid crystal device. Scientific Reports. 2015;**5**

[61] Farhadi S, Pourzare K, Bazgir S. Co_3O_4 nanoplates: Synthesis, characterization and study of optical and magnetic properties. Journal of Alloys and Compounds. 2014;**587**:632-637

[62] Farhadi S, Safabakhsh J, Zaringhadam P. Synthesis, characterization, and investigation of optical and magnetic properties of cobalt oxide (Co_3O_4) nanoparticles. Journal of Nanostructure in Chemistry. 2013;**3**(1):69

[63] Gürlük M et al. Comparison of guest-host liquid crystal systems doped with azo or anthraquinone dyes. Iranian Journal of Science and Technology, Transactions A: Science. 2017;**41**(1):1-5

Chapter 5

Substitution and Redox Properties of Some Organoisocyanide Cobalt(II) Complexes

Olayinka A. Oyetunji, Godiraone Tatolo,
Banyaladzi D. Paphane and Clifford A. L. Becker

Additional information is available at the end of the chapter

http://dx.doi.org/10.5772/intechopen.71448

Abstract

The reactions of four *tetrakis*(arylisocyanide)cobalt(II) complexes, $[Co(CNR)_4(ClO_4)_2]$ {R = 2,6-$Me_2C_6H_3$ (**A**), 2,4,6-$Me_3C_6H_2$ (**B**), 2,6-$Et_2C_6H_3$ (**C**), and 2,6-$iPr_2C_6H_3$ (**D**)}, with two pyridines, 4-CNpy and 4-Mepy, have been kinetically studied in trifluoroethanol medium. Each of the reactions, which was monitored over a temperature range of 293 to 318 K, exhibited two distinct processes proposed to be an initial fast substitution process followed by a slow reduction process. For each pyridine, steric hindrance was observed to play a significant role in the rates of the reactions, which decrease with increasing size of the arylisocyanide ligand in the order k(**A**) > k(**B**) > k(**C**) > k(**D**). Addition of each of three triarylphosphines, PR_3 (R = Ph, C_6H_4Me-p, C_6H_4OMe-p), to solutions of *pentakis* (t-octylisocyanide)cobalt(II), $[Co(CNC_8H_{17}\text{-}t)_5](ClO_4)_2$, resulted in a shift in the λ_{max} of the electronic spectrum accompanied by a change in color of the solutions. The shift is attributed to ligand substitution. The reactions of the cobalt(II) complex $[Co(CNC_8H_{17}\text{-}t)_5]^{2+}$ with the triarylphosphines are proposed to proceed via a combination of substitution, reduction, and disproportionation mechanisms with final formation of disubstituted Co (I) complexes. The order of reactivity of the complex with the triarylphosphines was found to be $P(C_6H_4OMe\text{-}p)_3$ > $P(C_6H_4Me\text{-}p)_3$ > PPh_3. This order is explained in terms of the electron donating/π-acceptor properties of the phosphines.

Keywords: substitution, redox, organoisocyanide, cobalt(II), kinetics, mechanisms

1. Introduction

Although isocyanides (C≡N-R) are versatile ligands in organometallic chemistry, their roles and applications in chemical systems have always been overshadowed by those of the CO ligand [1, 2]. Isocyanides show stronger σ-donor and weaker π-acceptor character which can be tuned by the substituent R and the complex fragment to which they are coordinated. The metal fragment has a substantial influence on the reactivity and stability of coordinated isocyanide ligands. Isocyanides form metal-ligand π-bonds by the back donation of electrons from the filled dπ orbitals of the metal center to the empty π* orbitals of the isocyanide ligand. Phosphine ligands on their part are phosphorus based, and they exhibit synergic interactions when they bond to metal centers. Phosphines (PR_3) utilize the *d*-orbitals to accept π-electron density. A number of techniques including infrared, ^{13}C and ^{31}P NMR, molecular mechanics, and reaction chemistry have been used to assess the binding capabilities of phosphorus ligands [3]. The predominant bonding ability of trialkyl phosphines is certainly from their very strong electron donating property, while π*-acceptance appears to be more significant for triaryl phosphines such as PPh_3. The synthesis and characterization of many cobalt (I), (II), and (III) complexes containing alkyl-, phenyl-, or aryl-isocyanide ligands have been reported [4–10].

Early studies on alkylisocyanide and arylisocyanide cobalt(II) complexes dating back to the sixties showed that while some of these complexes were reported as cobalt(II) dimers, a number of monomeric pentacoordinate cobalt(II) complexes were also reported [6, 11]. Reactions of cobalt(II)alkylisocyanide complexes with triarylphosphines have been studied and characterized as ligand substitution/reduction [10] with most reactions giving corresponding disubstituted five coordinate cobalt(I) complexes as final products, while the reactions of the cobalt(II) complexes with triarylarsines resulted in initial substitution, which at least in one case was accompanied by oxidation of the ligand to triarylarsine oxide and that of the cobalt (II) to cobalt(III) [12]. Spectra studies on these complexes suggested that some of them produce tetragonal structures in the solid state, while solution decomposition leads to tetrahedral geometry.

The five-coordinate alkylisocyanide complexes were known to generally react with tertiary phosphines to produce reduction/ligand substitution products of the type $[Co(CNR)_3(PR'_3)_2]X$, X = ClO_4 or BF_4 [13–15], exceptions being reactions with tri-n-alkylphosphines, which led to disproportionation/ligand substitution products of the types $[Co(CNR)_3(PR'_3)_2]X$ and $[Co(CNR)_3(PR'_3)_2]X_3$, X = ClO_4, BF_4 [13–15]. On the other hand, the five-coordinate arylisocyanide cobalt(II) complexes can react with tertiary phosphines to give monosubstituted cobalt(I) products of the type $[Co(CNR)_4(PR'_3)]X$ or di-substituted cobalt(I) products of the type $[Co(CNR)_3(PR'_3)_2]X$ depending principally on the level of steric hindrance in the arylisocyanide ligand [7, 16–20].

Conversely, triarylarsine ligands react with five-coordinate alkylisocyanide cobalt(II) complexes to produce primarily ligand-substituted six-coordinate Co(II) complexes of the type trans-$[Co(CNR)_4(AsR'_3)_2]X_2$, X = ClO_4, BF_4; PR'_3 = $AsPh_3$, $As(C_6H_4Me\text{-}p)_3$ [21]; although the Co(III) complex, $[Co(CNCH_2Ph)_4\{OAs(C_4H_4Me\text{-}P)_3\}_2](BF_4)_3$, has also been observed [12].

Apart from the many pentakis(arylisocyanide)cobalt(II) complexes that have been prepared, the synthesis and characterization of a number of tetrakis(arylisocyanide)bis(perchlorato) cobalt(II) complexes have also been reported [4, 22–24]. Earlier studies showed that these pentakis and tetrakis cobalt(II) complexes react with amines to undergo reduction to their corresponding cobalt(I) complexes of the general formula $[Co(CNR)_5]X$. In some cases, when R was sufficiently bulky, e.g. R = $CHMe_2$, CMe_3 and $C_6H_3iPr_2$, six-coordinate disubstituted intermediates like $[Co(CNR)_4(py)_2](ClO_4)_2$ were isolated [25–27]. It has also been observed that arylisocyanide cobalt(II) complexes are reduced in the presence of free arylisocyanide ligands or coordinating anions [6].

Reactions of cobalt(II) perchlorate hexahydrate with t-octylisocyanide, $CNCMe_2CH_2CMe_3$, in absolute ethanol are said to be different from known reactions of the same cobalt(II) salt with other alkylisocyanides but rather similar to reactions with arylisocyanide in which the monomeric five-coordinate complex is formed [28] according to the equation

$$Co(ClO_4)_2.6H_2O(al) + 5CNC_8H_{17}\text{-}t(al) \longrightarrow [Co(CNC_8H_{17}\text{-}t)_5](ClO_4)_2\ (s) \quad (1)$$

Unlike octahedral transition metal complexes, the formation, stability, and mechanistic behavior of five-coordinate square pyramidal or trigonal bipyramidal first row transition metal complexes have received significantly less attention. Such five-coordinate complexes can exhibit substantially different ligand substitution/redox behavior.

Some isocyano analogues of $[Co(CO)_4]$ in the −1, 0, and +1 oxidation states were synthesized a few years back [29] using m-terphenyl isocyanide as the stabilizing ligand. The same ligand was later employed in the synthesis of the isocyano counterpart of $HCo(CO)_4$ followed by a study of its decomposition and catalytic hydrogenation [30]. Figueroa and co-workers [31] most recently synthesized some isocyanide palladium(0) complexes as catalysts for the Suzuki-Miyaura cross-coupling of aryl bromides and arylboronic acids.

Despite the extensive information on the synthesis and characterization of these organo-isocyanide cobalt(II) complexes, little is known about kinetic studies on their substitution/reduction reactions. This chapter reports the kinetic studies on the substitution and redox reactions of two pyridines with some six-coordinate *tetrakis*(arylisocyanide)cobalt(II) complexes as well as those of five-coordinate *pentakis*(t-octylisocyanide)cobalt(II) complex with three triarylphosphines.

2. Experimental

2.1. Materials and syntheses of complexes

RNCHO, R = 2,6-$Et_2C_6H_3$, 2,6 $Me_2C_6H_3$, 2,6-$iPr_2C_6H_3$ and 2,4,6-$Me_3C_6H_2$, were prepared from commercial RNH_2 (Aldrich) and formic acid as described by McKusick and Webster [32]. These formamides were then used in the syntheses of the organoisocyanides. $Co(ClO_4)_2.6H_2O$ was obtained commercially from Strem Chemicals. Anhydrous diethyl ether was filtered through an alumina column before use and the *tetrakis*(arylisocyanide)cobalt(II) complexes, $[Co(CNC_6H_3Me_2\text{-}2,6)_4(ClO_4)_2]$ (**A**), $[Co(CNC_6H_2Me_3\text{–}2,4,6)_4(ClO_4)_2]$ (**B**), $[Co(CNC_6H_3Et_2\text{-}2,6)_4(ClO_4)_2]$ (**C**), and $[Co(CNC_6H_3iPr_2\text{–}2,6)_4(ClO_4)_2]$ (**D**) were prepared as reported in literature [4, 22, 24, 33]. Analar grade pyridines (Rochelle Chemicals) and trifluoroethanol (Fluka Chemicals) were used without further purification. *Pentakis*(t-octylisocyanide)cobalt(II) perchlorate, $[Co(CNC_8H_{17}\text{-}t)_5](ClO_4)_2$, was prepared as reported in the literature [28]. The yield was 96%. Characterization of the complexes was carried out using infrared and electronic spectra as well as elemental analyses. The results are in close agreement with those of earlier workers [4, 22, 24, 28, 33]. The ligands 1,1,3,3-tetramethylbutylisocyanide (*t*- octylisocyanide), $CNC_8H_{17}\text{-}t$, triphenylphosphine, PPh_3, tris(4-methoxyphenyl)phosphine, $P(C_6H_4OMe\text{-}p)_3$, tri (*p*-tolyl)phosphine, and $P(C_6H_4Me\text{-p})_3$ were purchased from Aldrich Chemicals and used without further purification. Absolute ethanol was obtained from Rochelle, and Analar grade dichloromethane was obtained from Saar Chem Chemicals. All other reagents used were of Analar grade.

2.2. Instrumentation

IR spectra were recorded on a Perkin Elmer 2000 FTIR spectrophotometer over the range 4000 to 400 cm^{-1}. Samples of the tetrakis(arylisocyanide)cobalt(II) complexes were prepared and run as Nujol mulls on NaCl plates or solution spectra over the range 3000 to 1000 cm^{-1}. The solutions from the reactions of Pyridine with the complexes in CF_3CH_2OH were used in their original state after the reaction was complete. Pyridine-CF_3CH_2OH mixture was used for background subtraction. Solution electronic spectra were recorded on a Shimadzu UV-2501PC spectrophotometer equipped with a Peltier TC 240A temperature regulator attachment over the range 1100–200 nm. Carbon, Hydrogen, and Nitrogen elemental analyses were performed on a Vario EL CHNOS Elemental Analyzer. Kinetic measurements were obtained from the Shimadzu spectrophotometer for the relatively slow reactions and a Hi-tech scientific SF-61 DX2 single mixing stopped-flow spectrophotometer for the fast reactions.

2.3. Kinetics

Kinetic measurements were taken on the Shimadzu UV-Vis spectrophotometer mentioned earlier for the slow reactions as well as on the stopped-flow spectrophotometer for the fast reactions. The cell compartments for both instruments were well thermostatted for constant temperature measurements between the temperatures of 293 and 318 K. This was done using the TC-240A temperature controller with the UV-Vis spectrophotometer, while water was circulated from a Neslab RTE 7 thermocirculator to the cell compartment of the stopped-flow equipment. Prior to kinetic measurements, the stabilities of the *tetrakis* cobalt(II) complexes were investigated in various solvents and were all found to be most stable in trifluoroethanol (CF_3CH_2OH). Absorbance changes monitored at the electronic absorption peak of the cobalt (II) complexes were found to remain practically unchanged in trifluoroethanol. Variation in the concentration of the trifluoroethanol did not affect the absorption spectra and the kinetics. For the reactions involving the *pentakis*(t-octylisocyanide) cobalt(II) complex, dichloromethane (CH_2Cl_2) was chosen as the appropriate solvent for the reactions as both the cobalt(II) complex and the triarylphosphines used (i.e., PPh_3, $P(C_6H_4OMe\text{-}p)_3$, and $P(C_6H_4Me\text{-p})_3$) completely dissolve in it. $[Co(CNC_8H_{17}\text{-}t)_5]$ $(ClO_4)_2$ was also found to be stable in CH_2Cl_2.

The kinetics were followed under pseudo-first-order conditions with the pyridines in concentrations which were 10 times or more greater than those of the *tetrakis* Co(II) complexes, and the triarylphosphines also had concentrations which were similarly 10 times or more greater than that of the *pentakis* cobalt(II) complex. All kinetic studies were followed to more than 90% completion. The reactions of the *tetrakis* complexes with the pyridines show an initial fast increase in absorbance followed by a slow gradual absorbance decrease. Conversely, the reaction between the *pentakis t-octyl* complex showed complex kinetics, while those of the same *pentakis* complex with the other two phosphines showed single exponential decay curves. For the slow processes, the pseudo first-order rate constants (k_s) were obtained from linear regression plots of $\ln(A_t\text{-}A_\infty)$ against time, t; where A_t and A_∞ are the absorbances at time t and $t = \infty$, respectively. Similar k_s values for the fast processes were obtained from the stopped-flow instrument kinetic software (KinetAsyst™ 3) by fitting the decay curves to single exponential analysis. These k_s values for the fast processes are averages of four to five runs for each pyridine concentration.

3. Conclusion

3.1. Kinetic studies

3.1.1. *Reactions of the four tetrakis(arylisocyanide)cobalt(II) complexes* **A, B, C,** *and* **D** *with pyridines*

All absorbance decay curves obtained while monitoring the reactions between the pyridines and the *tetrakis* cobalt(II) complexes show an initial fast increase in absorbance attributed to the substitution process, followed by a slow absorbance decrease representing the reduction process. **Figure 1** shows a typical decay curve for the reaction between $[Co(CNC_6H_3Et_2\text{–}2,6)_4(ClO_4)_2]$ and 4-cyanopyridine (4-CNpy).

It was also noticed that on addition of each pyridine, an immediate change from green to a nearly colorless solution (substitution) followed by slow development of a deep yellow solution (reduction), which persisted at the end of the reactions was observed. Repeated scans of the visible spectra on addition of each pyridine also showed a bathochromic shift of d-d

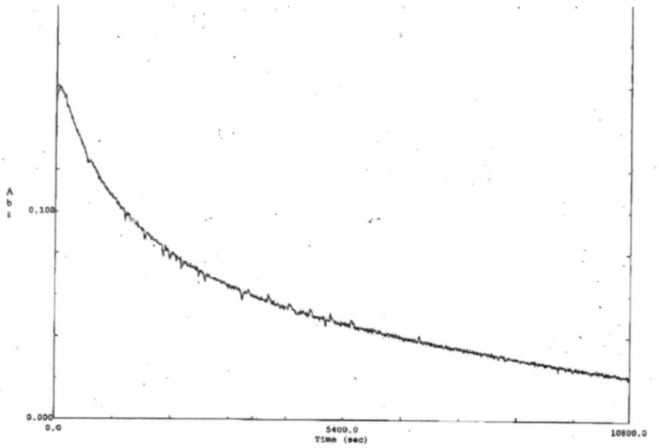

Figure 1. Typical absorbance decay curve for the reduction of $[Co(CNC_6H_3Et_2)_4(ClO_4)_2]$ complex by 4-CNpy at 298 K.

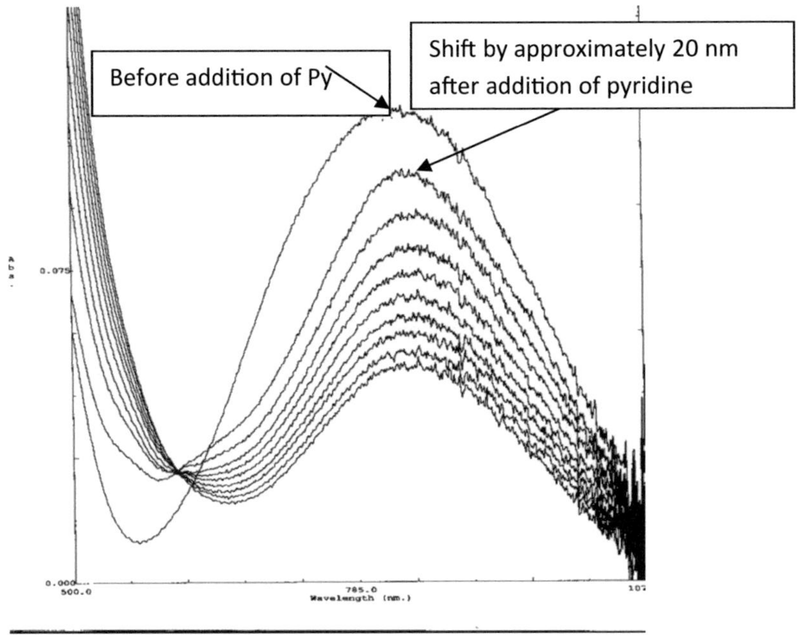

Figure 2. Spectral changes of a [$Co(CNC_6H_4Et_2\text{-}2,6)_4(ClO_4)_2$] solution in trifluoroethanol on addition of pyridine at 298 K.

transition by about 20 to 35 nm (**Figure 2**). This shift can be attributed to the substitution of the solvent molecules from the Co(II) inner shell by the incoming pyridine ligand (Eq. 2).

$$[Co(CNR)_4(sol)]^{2+} \xrightarrow{excessPy} [Co(CNR)_4(Py)_2]^{2+} \quad (2)$$

Similar observations have been reported by Becker [8, 21] in his work on the synthesis of some organoisocyanide cobalt(II) complexes as well as by other authors [34–37] in their investigations on reactions of cobalt(II) protoporphyrin complexes with amine ligands. The reduction of the *tetrakis*(arylisocyanide)cobalt(II) complexes by the pyridines was found to produce *pentakis*(arylisocyanide)cobalt(I) perchlorate salts as the final products. The isolation and characterization of some of these cobalt(I) complexes have been reported in earlier studies [38–40].

As explained earlier, the fast process was followed using stopped-flow techniques, while the slow process was followed using the conventional ultraviolet-visible spectrophotometry. Linear regression plots were used for the determination of the pseudo first-order rate constants. For convenience, we use k_{1s} for the fast substitution reactions and k_{2s} for the slow reduction processes. The reactions were monitored at different concentrations of the pyridines. For all the reactions, k_{1s} and k_{2s} increased with increasing concentration of pyridine. Plots of k_{1s} versus concentrations of pyridine were linear with intercepts. Using the data in **Tables 1–4**, such plots for the reactions of the four cobalt(II) complexes with 4-CNpy are shown in **Figure 3**.

For pyridine, the corresponding k_1 and k_2 values for the substitution process have been reported earlier [24] and are shown in **Table 4**:

10^3[4-CNpy], M	k_{1s}, s^{-1}			
	A	B	C	D
3.30	21.9	18.9	16.4	12.2
4.63	23.8	21.0	18.2	12.9
6.61	27.1	23.2	19.7	15.7
7.60	28.7	24.6	21.3	16.3
9.25	30.9	27.0	23.1	17.6
12.14	38.6	32.9	27.5	20.4
15.18	40.2	34.3	29.8	24.5
19.80	48.1	40.7	35.1	28.1
k_1, s^{-1}	16.6	14.8	12.7	8.7
k_2, M^{-1} s^{-1}	1610	1331	1143	992

Table 1. Pseudo–first-order rate constants, k_{1s}, for the substitution reactions between the four $[Co(CNR)_4(ClO_4)_2]$ complexes and 4-CNpy in CF_3CH_2OH at 298 K.

10^4[4-Mepy], M	k_{1s}, s^{-1}			
	A	B	C	D
3.30	29.9	27.3	23.6	20.8
4.63	34.2	30.8	26.5	23.3
6.61	39.1	34.2	29.5	26.0
7.60	43.1	36.1	31.6	27.9
9.25	47.6	39.8	34.2	30.1
12.14	55.1	45.0	39.3	34.2
15.18	66.0	52.6	43.6	37.5
19.80	78.9	61.5	52.0	44.9
19.80	2.59	1.33	0.35	0.162
k_1, s^{-1}	19.2	17.1	15.1	8.3
k_2, M^{-1} s^{-1}	1720	1550	1430	1030
k_1, s^{-1}	21.0	20.6	18.4	16.6
k_2, M^{-1} s^{-1}	29,751	20,652	16,909	14,213

Table 2. Pseudo–first-order rate constants, k_{1s}, for the substitution reactions between the four $[Co(CNR)_4(ClO_4)_2]$ complexes and 4-Mepy in CF_3CH_2OH at 298 K.

10^3[4-CNpy], M	$10^5 k_s'$, s^{-1}			
	A	B	C	D
3.30	1.23	0.67	0.24	0.079
4.63	1.44	0.85	0.27	0.094
6.61	1.90	0.94	0.30	0.106
7.60	1.98	1.08	0.31	0.112
9.25	2.20	1.13	0.33	0.131
12.14	2.27	1.24	0.34	0.138
15.18	2.43	1.29	0.35	0.150
K, M^{-1}	164.2	195.3	260.9	209.2
k_3, s^{-1}	3.49	1.74	0.398	0.198

Table 3. Pseudo–first-order rate constants, k_s', for the reduction of $[Co(CNR)_4(ClO_4)_2]$ complexes by 4-CNpy in CF_3CH_2OH at 318 K.

	A	B	C	D
k_1, s^{-1}	19.2	17.1	15.1	8.3
k_2, M^{-1} s^{-1}	1720	1550	1430	1030

Table 4. k_1 and k_2 values for the substitution reaction between pyridine and the four cobalt(II) complexes at 298 K.

The data fit well to the rate law described by the following equation:

$$\text{Rate} = \frac{-d[Co(CNR)_4^{\ 2+}]}{dt} = k_{1s}[Co(CNR)_4^{\ 2+}] \qquad (3)$$

This is consistent with octahedral substitution involving a two-step ligand-ligand replacement, according to the following mechanism:

$$[Co(CNR)_4(ClO_4)_2] + L \xrightarrow{k_1} [Co(CNR)_4(L)(ClO_4)]^+ + ClO_4^- \quad \text{slow} \qquad (4)$$

$$[Co(CNR)_4(L)(ClO_4)]^+ + L \xrightarrow{k_2} [Co(CNR)_4L_2]^{2+} + ClO_4^- \quad \text{fast} \qquad (5)$$

where k_{1s} in Eq. 3 above is given by.

$k_{1s} = k_1 + k_2[L]$.

From plots of k_{1s} vs. $[L]$, values of k_1 and k_2 were obtained from the intercepts and slopes, respectively, and are listed in **Table 1** and **2**. The fact that k_2 is much greater than k_1 indicates attack by the solvent is a much slower process than attack by the nucleophile.

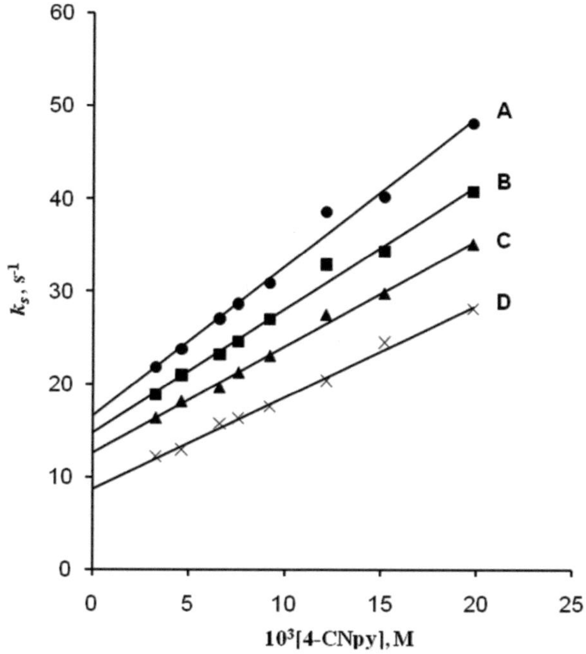

Figure 3. Plots of observed rate constants (k_{1s}) *versus* 4-CNpy concentration for the reaction between $[Co(CNR)_4(ClO_4)_2]$ complexes and 4-CNpy at 298 K.

In contrast, the slow reduction processes exhibited different kinetic behavior as compared to the substitution process. Plots of k_s' versus *[L]* gave curves typical of saturation kinetics (**Figure 4**). Similar observations have been reported in our earlier studies [24] as well as in both the aminolysis of sulfamate esters in chloroform [41] and in the reactions of Co(II) protoporphyrins IX dimethyl ester with pyridine and related compounds [34, 37]. Such saturation kinetics is also typical of the formation of a precursor complex prior to electron transfer as observed in the electron transfer reactions of halopentacyanocobaltate(III) complexes [42]. The above observation fits well with the following mechanism:

$$[Co(CNR)_4L_2]^{2+} \underset{\text{fast}}{\overset{K}{\rightleftharpoons}} [Co(CNR)_3L_2]^{2+} \quad + \ :CNR \xrightarrow[\text{slow}]{k_3} \text{Co(I) products} \tag{6}$$

giving the rate law:

$$\text{Rate} = k_s' [Co(CNR)_4L_2]^{2+} \quad \text{where } k_s' = \frac{k_3 K[L]}{1 + K[L]} \tag{7}$$

Rearrangement of Eq. 5 gives

$$\frac{1}{k_s'} = \frac{1}{k_3 K[L]} + \frac{1}{k_3} \tag{8}$$

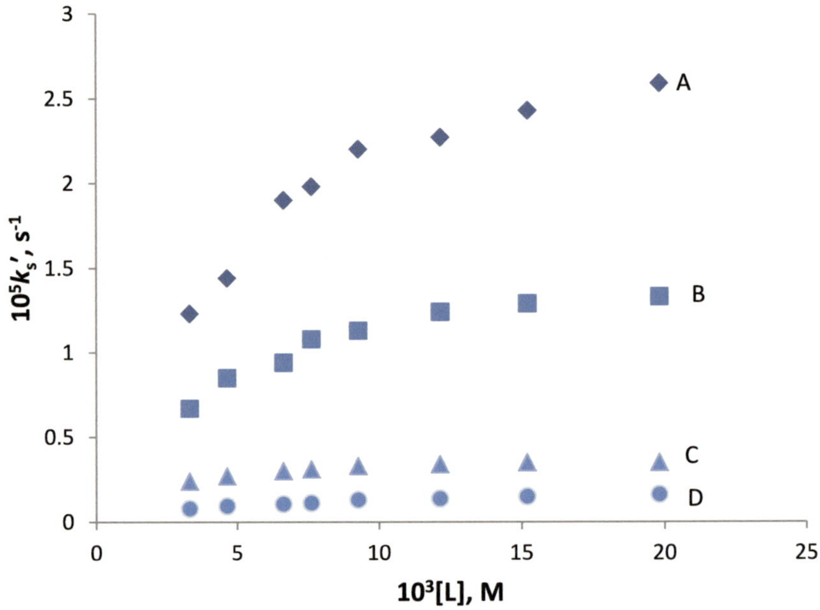

Figure 4. Plots of k_s' against concentration of 4-CNpy for the reduction of the four cobalt(II) complexes at 298 K.

Values of k_3 and K are reported in **Table 3**. It is observed from **Tables 1** and **3** that the rate constants k_2 and k_3 associated with the substitution by the amine ligand and the reduction of Co(II) to Co(I), respectively, decrease with increasing size of the arylisocyanide ligand, a clear evidence of steric hindrance.

For the reactions of the three pyridines under similar conditions, our results show that while 4-CNpy reacted at a slower rate than pyridine, 4-Mepy reacted at a faster rate. This observation can be explained in terms of the basicity of the amines. The CN group is an electron withdrawing substituent and therefore reduces the electron density of the pyridine ring, making 4-CNpy a weaker electron donor and hence less basic than pyridine. On the other hand, the methyl substituent on pyridine increases the electron density of the ring such that 4-Mepy is a better electron donor and a more basic amine than pyridine. Since ligands usually coordinate to metal centers by acting as Lewis bases donating lone pairs of electrons to the empty metal d-orbitals, more basic amines will be better coordinating ligands and will form more stable complexes as compared to less basic amines. This can be used to explain the higher reactivity of 4-Mepy and lower reactivity of 4-CNpy when they are compared to pyridine.

3.1.2. Reaction of $[Co(CNC_8H_{17}\text{-}t)_5](ClO_4)_2$ complex with triphenylphosphine in dichloromethane

Addition of PPh_3 to solutions of the *pentakis* cobalt(II) complex in CH_2Cl_2 resulted in an immediate shift in the peaks, one from 724 nm to approximately 880 nm and another from 486 nm to approximately 450 nm with attendant decrease in absorbance at both new wavelengths as the reaction progressed. It should be noted that the solutions of the *pentakis* complex

http://dx.doi.org/10.5772/intechopen.71448

employed in this study are those that have been standing for about 24 hours. The shift in peak on addition of PPh_3 with the attendant decrease in absorbance is attributed to ligand substitution followed by reduction, an observation similar to that reported in earlier studies [24, 43–45]. **Figure 5** shows the repeated UV/Vis spectra of the reaction mixture taken at intervals of time for 3 hours.

When the reaction of $[Co(CNC_8H_{17}\text{-}t)_5]^{2+}$ with PPh_3 was monitored at a fixed wavelength on the UV-VIS spectrophotometer over a period of 1 to 3 hours, some complex kinetics were observed as shown in **Figure 6**. Different attempts made to analyze the data did not yield any meaningful result. Repeated infrared scans were then employed to probe further insight into the mechanism of the reactions.

Stopped-flow kinetic trace for the substitution process between $[Co(CNC_8H_{17}\text{-}t)_5](ClO_4)_2$ and PPh_3 showed a perfect fit to a single exponential function (**Figure 7**), showing a first order dependence on the Co(II) complex.

The observed rate constants, k_s, were found to increase with concentration of PPh_3 and plots of k_s against $[PPh_3]$ were linear with nonzero intercepts, with a first-order dependence on triphenylphosphine. This result is consistent with the following rate law:

$$\text{Rate} = -\frac{d[Co(CNC_8H_{17}-t)_5]^{2+}}{dt} = ks[\{Co(CNC_8H_{17}-t)_5\}^{2+}] = (k_1 + k_2[PR_3])[\{Co(CNC_8H_{17}-t)_5\}^{2+}] \quad (9)$$

where $k_s = k_1 + k_2[PR_3]$

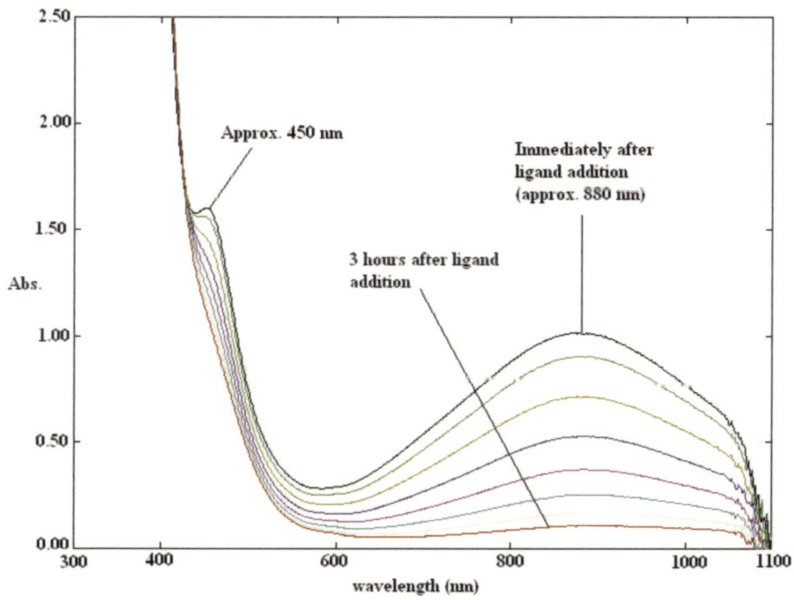

Figure 5. Typical repeated scans of the reaction of $[co(CNC_8H_{17}\text{-}t)_5]^{2+}$ with PPh_3 in CH_2Cl_2 at 298 K.

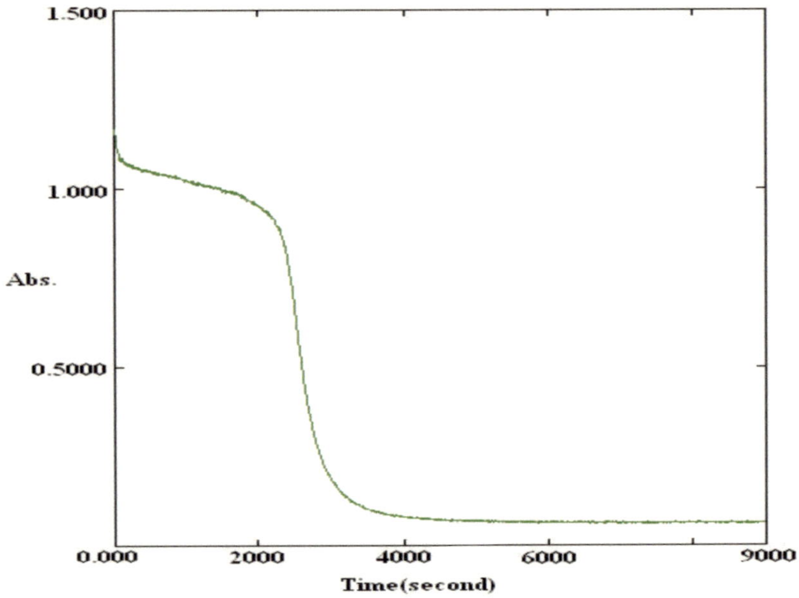

Figure 6. Typical absorbance decay curve for the reaction of $[Co(CNCHMe_2)_5]^{2+}$ with PPh_3 at 880 nm.

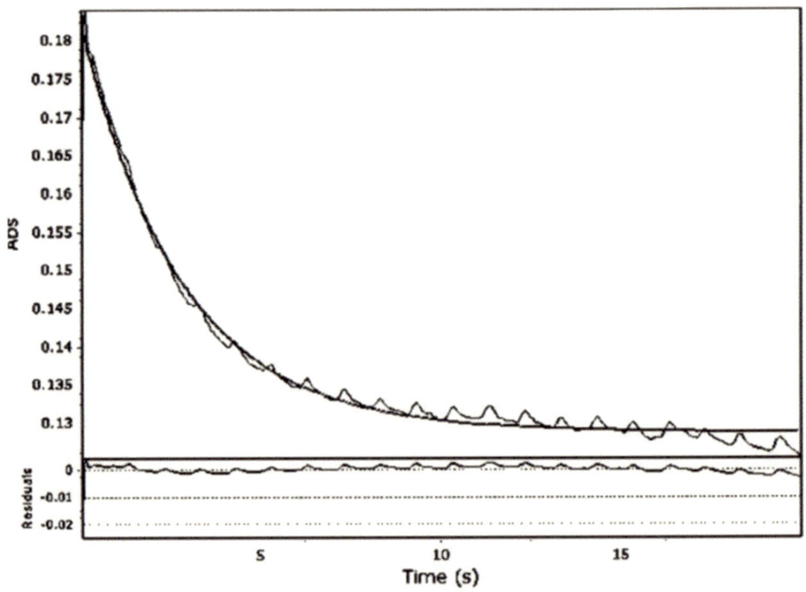

Figure 7. Typical kinetic trace for the reaction between $[Co(CNC_8H_{17}\text{-}t)_5]^{2+}$ and PPh_3 fitted to a single exponential function.

This is similar to the rate law reported for the substitution reaction between freshly prepared solution of $[Co(CNC_8H_{17}\text{-t})_5]^{2+}$ and PPh_3 where it was suggested that the substitution process involved a solvent-assisted replacement of one isocyanide ligand with PPh_3 [46]. However, the stale solutions in this study substitutes at slower rates. For example, while the second order rate constant for the substitution of the isocyanide ligand by PPh_3 in the freshly prepared solution at 298 K was 943 $M^{-1}\ s^{-1}$, the rate constant for the stale solution was 782 $M^{-1}\ s^{-1}$. It is suggested that the stale solutions contain solvent molecules in their coordination shells as a cobalt(II) intermediate before undergoing substitution of one isocyanide ligand by PPh_3.

Attempts to analyze the complex kinetic traces did not yield meaningful results, and therefore, infrared studies were employed in an effort to probe further insight into the mechanism of the reactions. The repeated infrared scan taken over a period of 3 hours for reactions of solutions of the Co(II) complex with PPh_3 show gradual disappearance of the Co(II) peak just as the Co(I) peak became more intense. Twelve hours after the reaction between the t-octyl Co(II) complex and PPh_3 started, the infrared spectra of the reaction mixture showed the isocyanide stretching frequencies largely associated with pentakis Co(I) isocyanide complexes as well as some peaks characteristic of mono- and di-substituted Co(II) and the starting Co(II) complex. This observation is an indication of a slow reaction between $[Co(CNC_8H_{17}\text{-t})_5]^{2+}$ and PPh_3 probably involving some multistep processes including pre-equilibria. This is similar to what was obtained with freshly prepared solutions of $[Co(CNC_8H_{17}\text{-t})_5]^{2+}$ and an indication of similar reaction mechanisms, which could include the following reaction pathways:

$$2[Co(CNR)_5]^{2+} + 3PR_3 \rightleftharpoons [Co(CNR)_4(PR_3)]^{2+} + [Co(CNR)_3(PR_3)_2]^{2+} + 3CNR \quad (10)$$

$$[Co(CNR)_4(PR_3)]^{2+} \longrightarrow [Co(CNR)_3(PR_3)_2]^{2+} + CNR \quad (11)$$

$$[Co(CNR)_3(PR_3)_2]^{2+} + e \longrightarrow [Co(CNR)_3(PR_3)_2]^{+} \qquad CNR = CNC_8H_{17}\text{-t}; \quad PR_3 = PPh_3 \quad (12)$$

3.1.3. *Reaction of $[Co(CNC_8H_{17}\text{-t})_5]^{2+}$ complex with tri(p-tolyl)phosphine and tris(4-methoxyphenyl) phosphine*

When $P(C_6H_4Me\text{-p})_3$ and $P(C_6H_4OMe\text{-}p)_3$ were separately added to stale solutions of the cobalt(II) complex, shifts in absorption peaks to higher wavelengths from 724 nm to approximately 955 nm for $P(C_6H_4Me\text{-p})_3$ and 724 nm to 904 nm for $P(C_6H_4OMe\text{-}p)_3$ were observed with attendant color changes. However, these reactions did not show complex kinetics exhibited by PPh_3 but show single experimental decays under similar experimental conditions. Although these reactions also involve different pre-equilibria and equilibria processes in addition to the reduction of Co(II) to Co(I), the substitution reactions must be occurring at a relatively fast process (too fast to be measured by the available instrumentation) that the single exponential decay observed can be attributed mainly to the reduction of the Co(II) complex to its Co(I) analog. As attempts to analyze the kinetic traces did not give reproducible results, the discussion of the reactions is mainly based on infrared studies.

Infrared spectra taken at different time intervals show that there was fast reduction of the Co(II) complex to Co(I) as strong peaks of the Co(I) complexes appear immediately after addition of these phosphines. Furthermore, the reactions of the two phosphines with stale solutions of the Co(II) complex also resulted in disproportionation reactions. This is evidenced by the observed infrared peaks due to the Co(III) and Co(I) analogues of the Co(II) complex with the di-substituted Co(III) complex, showing its characteristic isocyanide stretching frequency at 2234 cm^{-1} and its Co(I) analogue at 2063 cm^{-1} (**Figure 8**). Earlier studies had recorded similar observation in the reactions of some alkylisocyanide cobalt(II) complexes with tri-(p-tolyl)phosphine and 2-furylphosphine [47, 48].

Among the possible processes that take place in these reactions are:

$$[Co(CNR)_5]^{2+} + 2PR_3 \xrightarrow{\text{fast}} [Co(CNR)_4(PR_3)_2]^{2+} + CNR \quad (13)$$

$$[Co(CNR)_5]^{2+} + 2PR_3 \rightleftharpoons [Co(CNR)_3(PR_3)_2]^{2+} + 2CNR \quad (14)$$

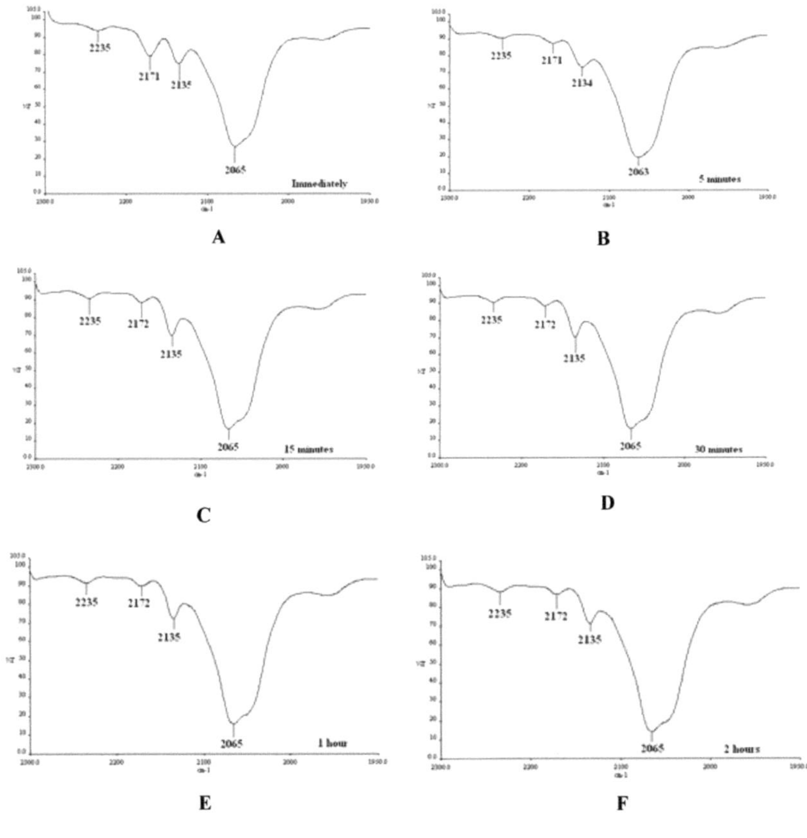

Figure 8. Isocyanide stretching frequencies for the reaction between stale $[Co(CNC_8H_{17}\text{-}t)_5$ and tris(4-methoxyphenyl) phosphine at different time intervals (A–F).

$$2[Co(CNR)_4(PR_3)_2]^{2+} \rightleftharpoons [Co(CNR)_4(PR_3)_2]^{3+} + [Co(CNR)_3(PR_3)_2]^{+} + CNR \quad (15)$$

$$[Co(CNR)_3(PR_3)_2]^{2+} + e \longrightarrow [Co(CNR)_3(PR_3)_2]^{+} \quad (16)$$

$$CNR = CNC_8H_{17}\text{-t} \quad \text{and } PR_3 = P(C_6H_4Me\text{-p})_3\ ;\ P(C_6H_4OMe\text{-}p)_3$$

Acknowledgements

The authors are grateful to University of Botswana Research Committee for funding.

Author details

Olayinka A. Oyetunji[1]*, Godiraone Tatolo[1], Banyaladzi D. Paphane[2] and Clifford A. L. Becker[3]

*Address all correspondence to: yoyetunji@gmail.com

1 Department of Chemistry, University of Botswana, Gaborone, Botswana

2 Department of Basic Sciences, Botswana University of Agriculture and Natural Resources, Gaborone, Botswana

3 Retired

References

[1] Han Y, Huynh HV. Dalton Transactions. 2009;(12):2201

[2] Slaughter LM. Inorganic Chemistry Communications. 2008;**29**:46

[3] Bodner GM. Inorganic Chemistry. 1975;**14**:2694

[4] Becker CAL. Synthesis and Reactivity in Inorganic and Metal-Organic Chemistry. 1992; **22**(4):437

[5] Becker CAL, Sebobi GS, Simane NT. Inorganica Chimica Acta. 2002;**334**:327

[6] Malatesta L, Bonati F. Isocyanide Complexes of Metals. New York, Chapter 7 and references therein: John Wiley; 1969

[7] Becker CAL. Synthesis and Reactivity in Inorganic and Metal-Organic Chemistry. 1979;**9**:445

[8] Becker CAL. Synthesis and Reactivity in Inorganic and Metal-Organic Chemistry. 1992;**22**:99

[9] Becker CAL, Motladiile S. Synthesis and Reactivity in Inorganic and Metal-Organic Chemistry. 2001;**31**(9):1545

[10] Becker CAL. Journal of Coordination Chemistry. 2000;**50**:89

[11] Cotton FA, Dunne TG, Wood JS. Inorganic Chemistry. 1964;**3**:1495

[12] Becker CAL. Inorganica Chimica Acta. 1993;**203**:175

[13] Becker CAL, Al-Qallaf SA, Cooper JC. Inorganica Chimica Acta. 1991;**188**:99

[14] Becker CAL, Biswas MAS, Cooper JC. Inorganica Chimica Acta. 1991;**188**:191

[15] Becker CAL, Biswas MAS. Journal of Coordination Chemistry. 1993;**29**:227

[16] Dart JW, Lloyd ML, Mason R, McCleverty JA, Williams J. Journal of the Chemical Society, Dalton Transactions. 1973:1747

[17] Bordignon E, Croatto U, Mazzi U, Orio AA. Inorganic Chemistry. 1974;**13**:935

[18] Otsuka S, Rossi M. Bulletin of the Chemical Society of Japan. 1973;**46**:3411

[19] Becker CAL. Journal of Inorganic and Nuclear Chemistry. 1980;**42**:27

[20] Becker CAL, Homsy NK. Inorganica Chimica Acta. 1984;**89**:93

[21] Becker CAL. Journal of Coordination Chemistry. 1994;**31**:337

[22] Becker CAL, Cooper JC. Inorganica Chimica Acta. 1989;**158**:141

[23] Becker CAL. Synthesis and Reactivity in Inorganic and Metal-Organic Chemistry. 1995;**25**:1455

[24] Oyetunji OA, Paphane BD, Becker CAL. Transition Metal Chemistry. 2006;**31**:951

[25] Becker CAL, Odisitse S. Synthesis and Reactivity in Inorganic and Metal-Organic Chemistry. 2000;**30**:1547

[26] Becker CAL. Journal of Coordination Chemistry. 1992;**26**:231

[27] Becker CAL, Biswas MAS. Synthesis and Reactivity in Inorganic and Metal-Organic Chemistry. 1995;**25**:269

[28] Becker CAL, Ramotadima O, Makono VB, Baipoledi KC, Ahmad J. Journal of Coordination Chemistry. 2008;**iFirst**:1

[29] Margulieux GW, Weidermann N, Lacy DC, Moore CE, Rheingold AL, Figueroa JS. Journal of American Chemical Society. 2010;**132**(14):5033

[30] Carpenter AE, Rheingold AL, Figueroa JS. Organometallics. 2016;**35**(14):2309

[31] Barnett BR, Labios LA, Stauber JM, Moore CE, Rheingold AL, Figueroa JS. Organometallics. 2017;**36**(4):944

[32] McKusick BC and Webster OW, Organic Syntheses, vol. 41, Wiley, New York, 1961, pg 102–103

[33] Becker CAL, Cooper JC. Inorganica Chimica Acta. 1991;**182**:25

[34] Pavlovic D, Asperger S, Domi B. Journal of the Chemical Society Dalton Transactions. 1986:2535

[35] Stynes DV, Stynes HC, James BR, Ibers JA. Journal of the American Chemical Society. 1973;**95**:1796

[36] Datta-Gupta N. Journal of Inorganic and Nuclear Chemistry. 1972;**33**:419

[37] Dokuzovic Z, Ahmeti X, Pavlovic D, Murati I, Asperger S. Inorganic Chemistry. 1982;**21**:1576

[38] Becker CAL, Davis BL. Journal of Inorganic and Nuclear Chemistry. 1977;**39**:781

[39] Becker CAL. Journal of Inorganic and Nuclear Chemistry. 1975;**37**:703

[40] Becker CAL. Journal of Coordination Chemistry. 1993;**29**:147

[41] Spillane WJ, Hogan G, McGrath P, King J. Journal of the Chemical Society Dalton Transactions. 1998;**2**:309

[42] Ojo J, Ige J, Ogunlusi G, Owoyomi O, Olaseni E. Transition Metal Chemistry. 2006;**31**:337

[43] King RB, Saran MS. Inorganic Chemistry. 1972;**11**(9):2112

[44] Becker CAL, Anisi A, Myer G, Wright JD. Inorganica Chimica Acta. 1986;**111**(1):11

[45] Albertin G, Bordignon E, Orio AA, Rizzardi G. Inorganic Chemistry. 1975;**14**(4):944

[46] Oyetunji OA, Ramokongwa G, Becker CAL. Journal of Coordination Chemistry. 2013;**66**:2489

[47] Becker CAL. Synthesis and Reactivity in Inorganic and Metal-Organic Chemistry. 1998; **28**(7):1145

[48] Sweigart DA, Cooper DE, Millican JM. Inorganic Chemistry. 1974;**13**:1272

Chapter 6

Effect of Cobalt on Morphology of Microhole Formed by Micropunching

Kelvii Wei Guo and Hon Yuen Tam

Additional information is available at the end of the chapter

http://dx.doi.org/10.5772/intechopen.70514

Abstract

Due to the trend of miniaturization, the specific advantages of metal forming, especially for the high productivity and material utilization, cannot be exploited to the same extent in the field of electronics production as in conventional metal forming. Because of the competition with less productive and environmentally more polluting technologies, micropunching is extremely suitable for mass production of micro features, and the quality of the relevant punched microholes is definitely attracted to its successful application. Herein, the effect of cobalt on the morphology of microhole formed by punching with tungsten carbide-cobalt (WC/Co) micropunches was investigated. The results reveal that due to the optimal joint-contribution of WC and Co, the morphologies of microholes punched by 75% vol. WC+ 25% vol. Co micropunches not only satisfy with the practical requirements but also the punch using life is at the highest level among the three types of micropunches. Moreover, for 75WC/25Co micropunch, the serious wear of micropunch occurs with the wear loss of Co and WC when the punching number exceeds 1525. With the further increase in punching numbers, the dominant factors of the wear loss would mainly rely on the easily peeled off WC due to the serious loss of Co.

Keywords: cobalt, micropunch, WC/Co, microhole, morphology

1. Introduction

The demand of miniaturization comes from electronic devices, medical equipment, sensor technology and optoelectronics. Microfabrication technologies play a crucial role in product miniaturization [1–3]. Punching is the process of forcing a punch through the material and into a die to create a hole in the work piece. Micropunching can be an economic way of fabricating shaped microholes in mass production. The ability to fabricate microholes in large quantities has potential applications in micro-chip packaging, ink-jet print-head manufacture, bio-chip technologies and so on. Work in micropunching was reported by many investigators [4–7].

However, tool wear is an important issue in micro punching. Even with punches made of hard and tough materials like tungsten carbide-cobalt (WC/Co), quality of the punched holes declines rapidly under repeated punching [7].

Therefore, the effect of cobalt on the morphology of microhole formed by punching pure titanium (Ti) with WC/Co micropunches was investigated. The morphology variation of the micropunched holes at various cobalt volumes was measured by scanning electron microscopy (SEM), energy dispersive spectrometer (EDS) and confocal laser to reveal the relationship between the morphology of punched microhole and the ratio of cobalt in the micropunch.

2. Experimental materials and procedures

2.1. Experimental material

Micropunches with 75, 50, 25% volume fraction of WC particle and 25, 50, 75% volume fraction of Co particle of 50 μm mean size, 150 μm in diameter, was fabricated, respectively. The typical profile of the fabricated micropunch is shown in **Figure 1**. Pure titanium sheet with 200 μm in thickness was used as the substrate.

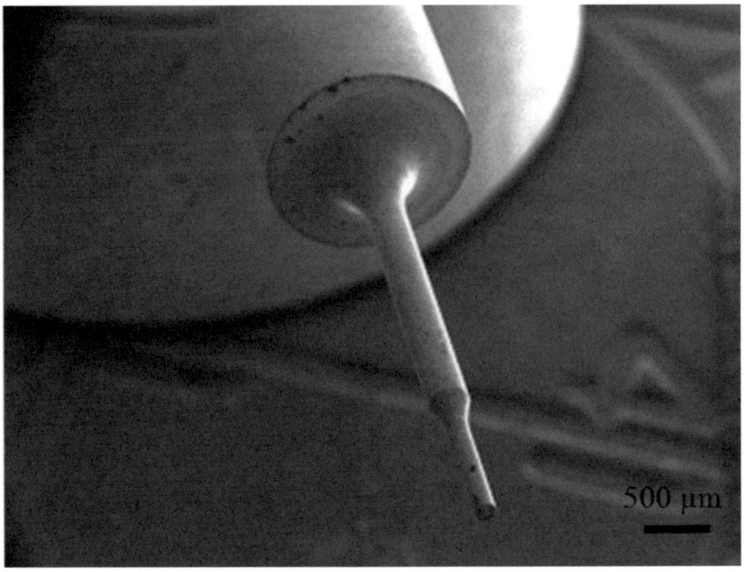

Figure 1. Profile of micropunch.

2.2. Experimental procedures

The prepared pure titanium sheet was properly cleaned by acetone and pure ethyl alcohol so as to remove any possible contaminant and carefully put into the microdie. Thereafter, specimens were punched by the microprocessing machine MP50 (made in Japan) with 20 pulses per minute and feedrate of 2 mm.

Punching experiments were done using an in-house setup. Precise vertical alignment during punching is crucial as lateral position error can lead to premature tool wear, tool breakage as well as inferior quality of the punched holes. The setup contained a vertical motion carriage with less than 1.5 μm lateral error over the 20 mm vertical stroke. The punch gripper that hosted the micropunch was fixed to the carriage. The stroke and load (maximum 20 N) of the carriage were provided by a z-stepper through a push rod. Specimens were held between a pair of 5 mm diameter bushes during the micropunching. The bush material was WC/Co, same as that of the micropunch. The bushes contained a 200 μm through-hole with 10° taper. The upper and lower faces of the bushes were carefully ground to ensure flatness and parallelism.

The effect of cobalt on the morphology of microhole at various micropunches was investigated by scanning electron microscopy (SEM) and energy dispersive spectrometer (EDS). The diameter of the machined microhole by micropunch was measured by LEXT confocal laser-OLS3000.

3. Results and discussion

3.1. Wear of micropunches with various cobalt ratios

In the micropunching, the wear of three types of micropunches in the initial is all significantly increased [7], and the dominant factor of the wear loss is mainly due to Co.

With the punching number increasing, the wear of WC/Co micropunch is in the quasistable period with a little wear loss.

Figure 2 illustrates its corresponding surface textures of three types of micropunches. It shows small pieces of WC particles distribute uniformly on the surface with 75% WC + 25% Co as shown in **Figure 2a**. While the volume fraction of Co increases, WC particles cannot observe distinctively, especially for the result shown in **Figure 2c**. It indicates that the fabricated micropunches with higher volume fraction of Co particles are not suitable for micropunching to meet with the desired requirements.

3.2. Effect of cobalt on morphology of punched microholes

The morphology of the punched microholes in the quasistable condition is shown in **Figure 3**. It illustrates that although there distributes some debris in the back side (**Figure 3a(ii)**, **b(ii)** and **c(ii)**), the profile of microhole formed with 75% WC + 25% Co is obviously better than that of the other two types (cf. **Figure 3a–c**).

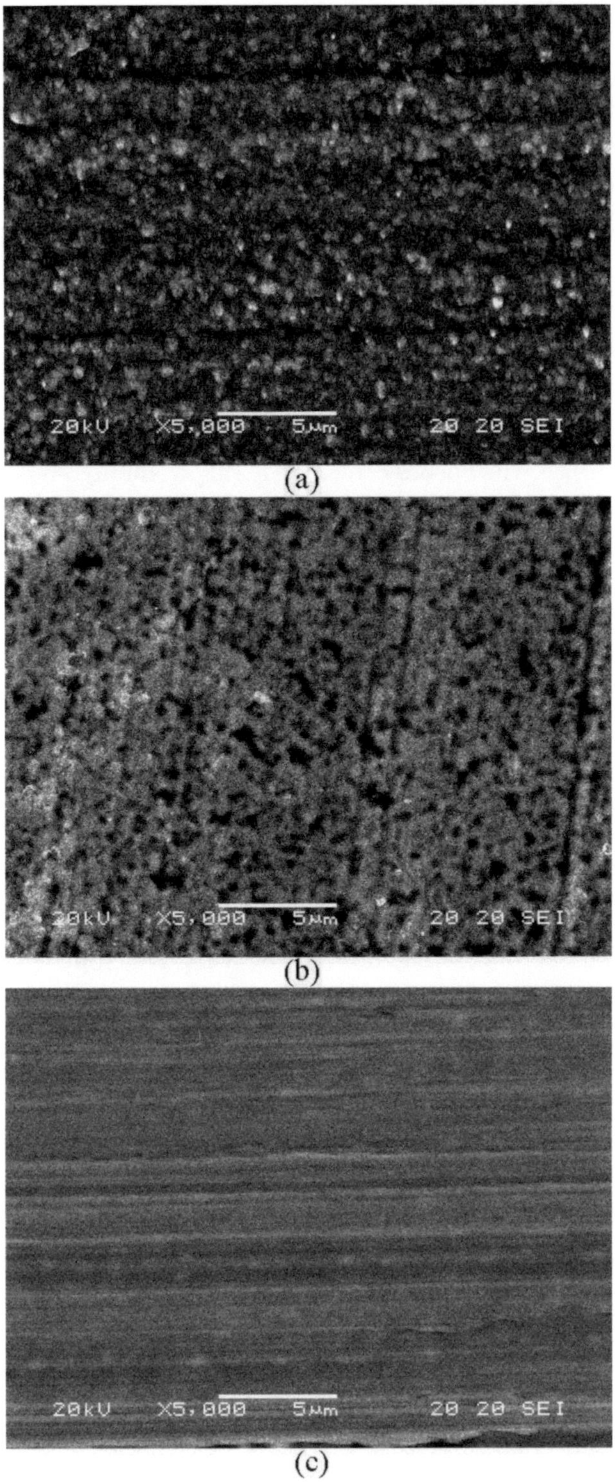

Figure 2. Surface texture of micropunches in the quasistable condition. (a) 75% WC + 25% Co. (b) 50% WC + 50% Co. (c) 25% WC + 75% Co.

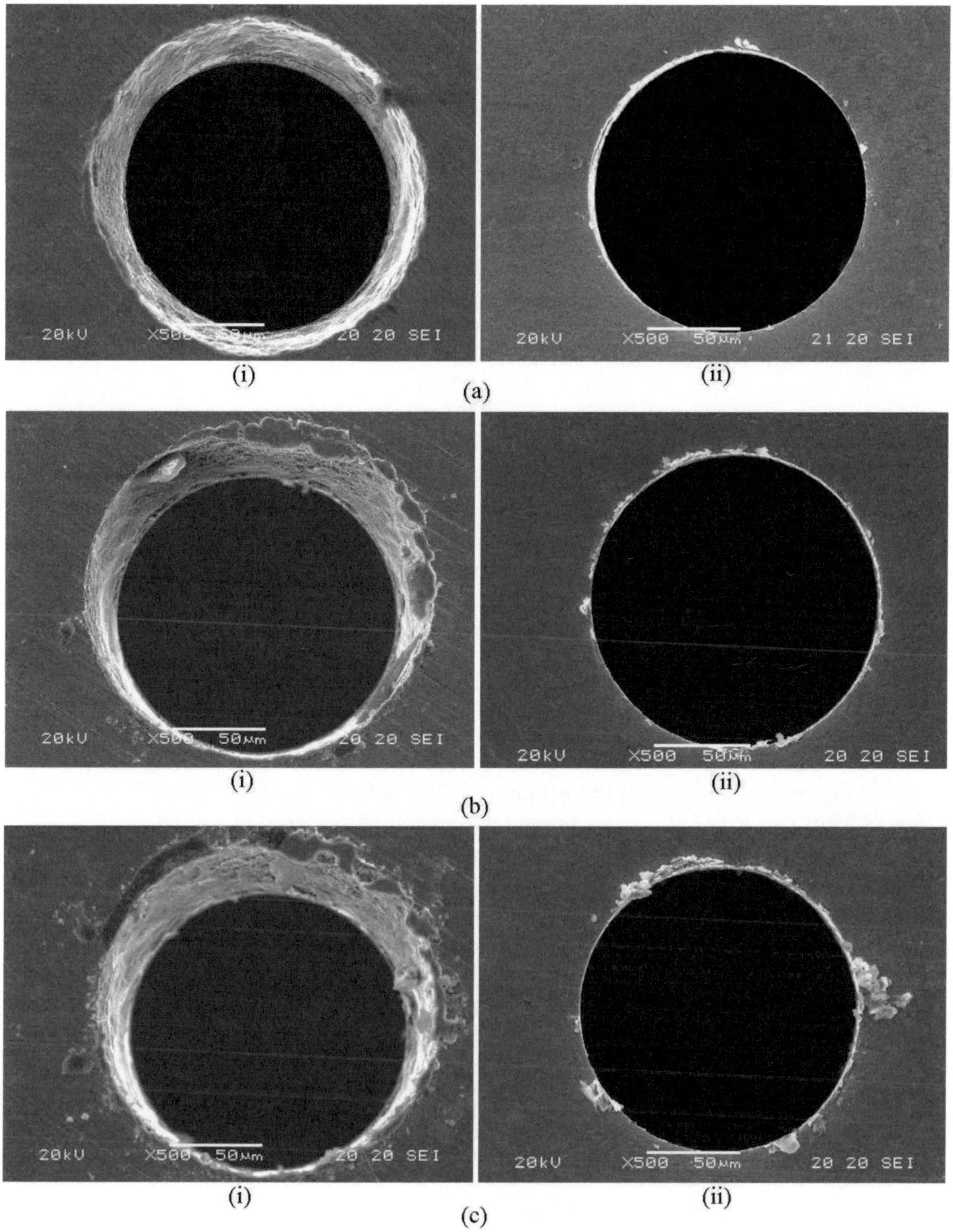

Figure 3. Morphology of punched microholes. (a) 75% WC + 25% Co. (b) 50% WC + 50% Co. (c) 25% WC + 75% Co.

Moreover, with the increment of Co, the debris distributed both on front and back sides increases. The morphology of the punched microhole is not even well, especially for the front side (**Figure 3b(i)** and **c(i)**). For the punches with 50% WC + 50% Co and 25% WC + 75% Co, the large bulk debris distributed on the surface results in the lower quality of the microhole, especially for the microholes punched with 25% WC + 75% Co.

It reveals that due to the optimal joint-contribution of WC and Co, i.e. 75% WC + 25% Co, the morphologies of the punched microholes not only satisfy with the practical requirements but also the wear loss of micropunch in the quasistable period is at the lowest level.

Furthermore, only small variation of the diameter of the microhole was experienced in the quasistable stage by the micropunch with 75% WC + 25% Co, as the weight loss was also small in the quasistable condition [7].

3.3. Effect of cobalt on life of micropunches

The life of types of micropunches is shown in **Figure 4**. It elucidates that the life of micropunch with 75% WC + 25% Co is definitely longer than the other two types, which agrees with the results shown in **Figure 2**. The effective punching number of the micropunch with 75% WC + 25% Co is about 1200. On the contrary, the life of micropunches with 50% WC + 50% Co is lower than 600. Even worse, for punches with 25% WC + 75% Co, its effective punching number is less than 100. Consequently, as an achievable tool for fabricating microholes, the micropunches with 50% WC + 50% Co and 25% WC + 75% Co are not capable for micropunching.

3.4. Quasistable wear characteristic of micropunch with 75% WC + 25% Co

The relationship between the wear loss 75WC/25Co micropunch and punching numbers is shown in **Figure 5**. It shows that the weight of 75WC/25Co micropunch (each for five times) has an obvious decrease with the increment of punching number in the initial.

After the obvious wear of 75WC/25Co micropunch in the initial and with the punching number increasing, the wear of 75WC/25Co micropunch is in the quasistable period with a little wear loss as shown in **Figure 5**, especially for punching number from 500 to 1200. **Figure 6** illustrates its corresponding surface texture of 75WC/25Co micropunch. It shows that small pieces of WC particles are observed on the surface. In the meantime, WC particles are distributed uniformly on the micropunch surface.

As comparison, the morphology of the punched microhole by 75WC/25Co micropunch in the initial period is expressed in **Figure 7**. Some substrate debris is distributed sparsely in the back side as shown in **Figure 7b**, and its EDS results are illustrated in **Figure 8**.

According to the morphology of the microhole punched by 75WC/25Co micropunch as shown in **Figure 3a**, it illustrates that although there still distributes some substrate debris in the back side, the profile of microhole formed in quasistable period is better than that of the initial period (cf. **Figures 3a** and **7**). It reveals that due to the joint-contribution of WC and Co with the optimal ratio (75% volume fraction WC particle + 25% volume fraction Co), the wear loss of micropunch in the quasistable period is the least among the three types of micropunches.

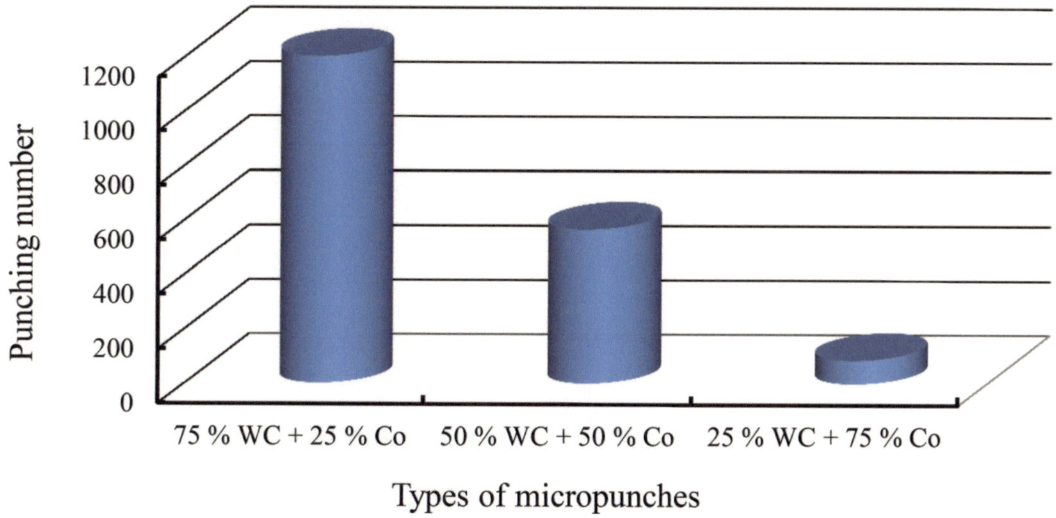

Figure 4. Life of different micropunches.

3.5. Profile of the microhole punched by 75WC/25Co micropunch

The diameter of the punched microhole by 75WC/25Co micropunch was measured by LEXT confocal laser-OLS3000 as shown in **Figure 9**. Its corresponding results (each for five times) are shown in **Figure 10**.

Compared with **Figure 5**, it illustrates that in the different wear conditions the diameter of microhole is changed correspondingly. In the initial condition, the diameter of microhole decreases obviously with the increment of punching number, which definitely agrees with results shown in **Figure 5**. While the punching number is from 500 to 1200, the diameter of microhole is relatively kept stable. With the punching number increasing further, especially when the punching number

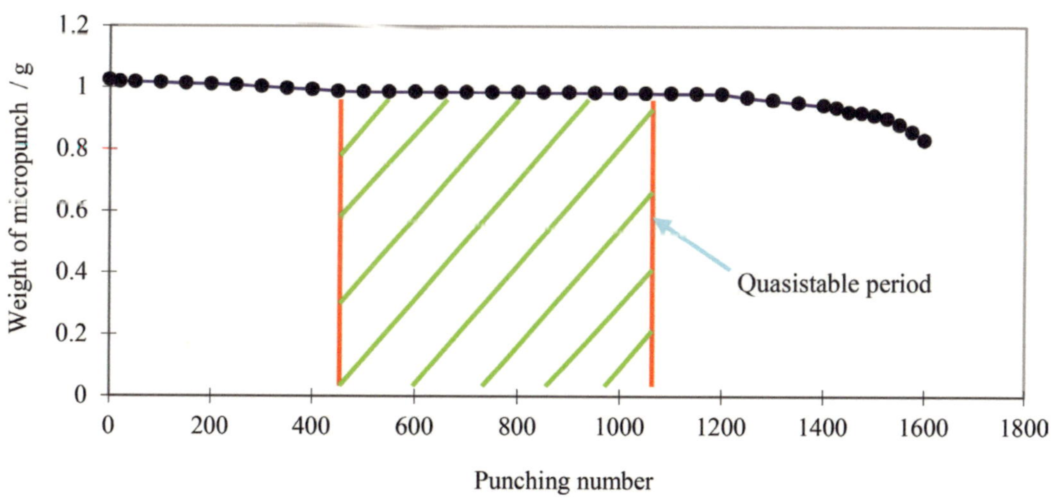

Figure 5. Relationship between wear loss of 75WC/25Co micropunch and punching numbers.

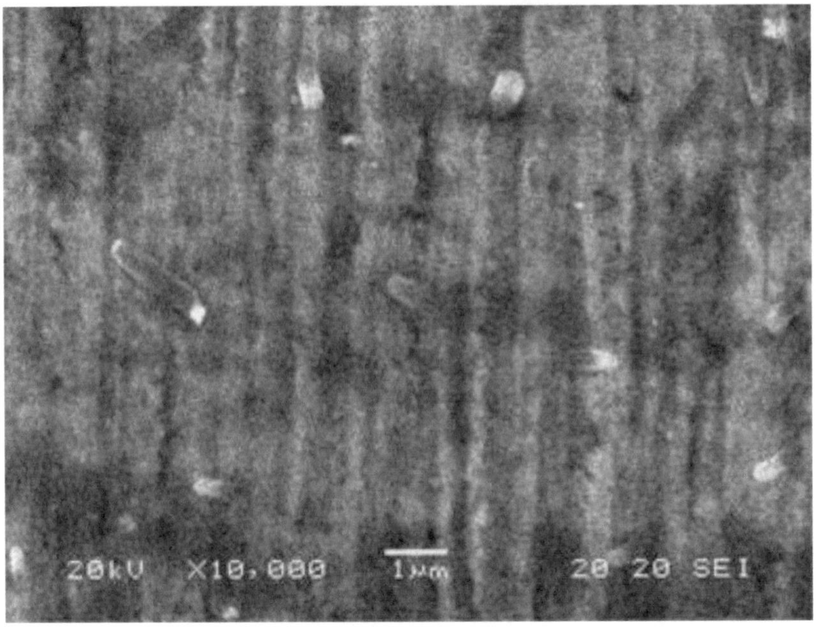

Figure 6. Surface texture of 75WC/25Co micropunch in the quasistable condition (punching number between 500 and 1200).

exceeds 1525 as shown in **Figure 10**, the diameter of microhole decreases remarkably because of the serious loss of Co. Consequently, the serious wear of micropunch mainly relies on WC particles which are easily peeled off without Co as bonding material in the severe wear condition of micropunching. Meanwhile, due to the temperature increment with punching number increasing, WC particles are more easily peeled off resulting in more intensive wear loss. Moreover, if the punch number increases in a minute, the wear loss of micropunch will be more drastically. In addition to the abovementioned, the microstructure of micropunch should be considered further, especially for the distribution of WC and Co particles.

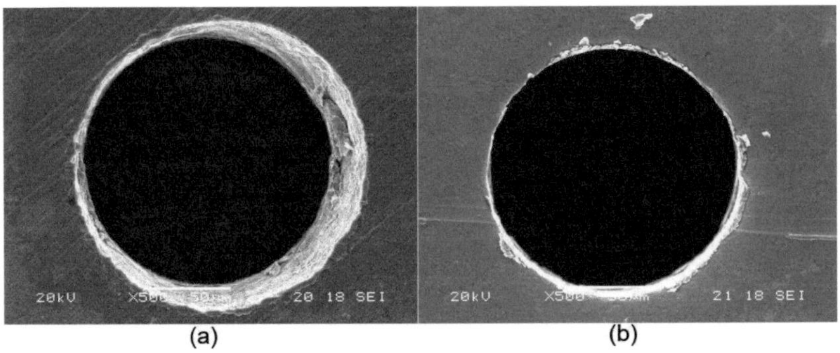

Figure 7. Morphology of microhole punched by 75WC/25Co micropunch in the initial period. (a) Front side. (b) Back side.

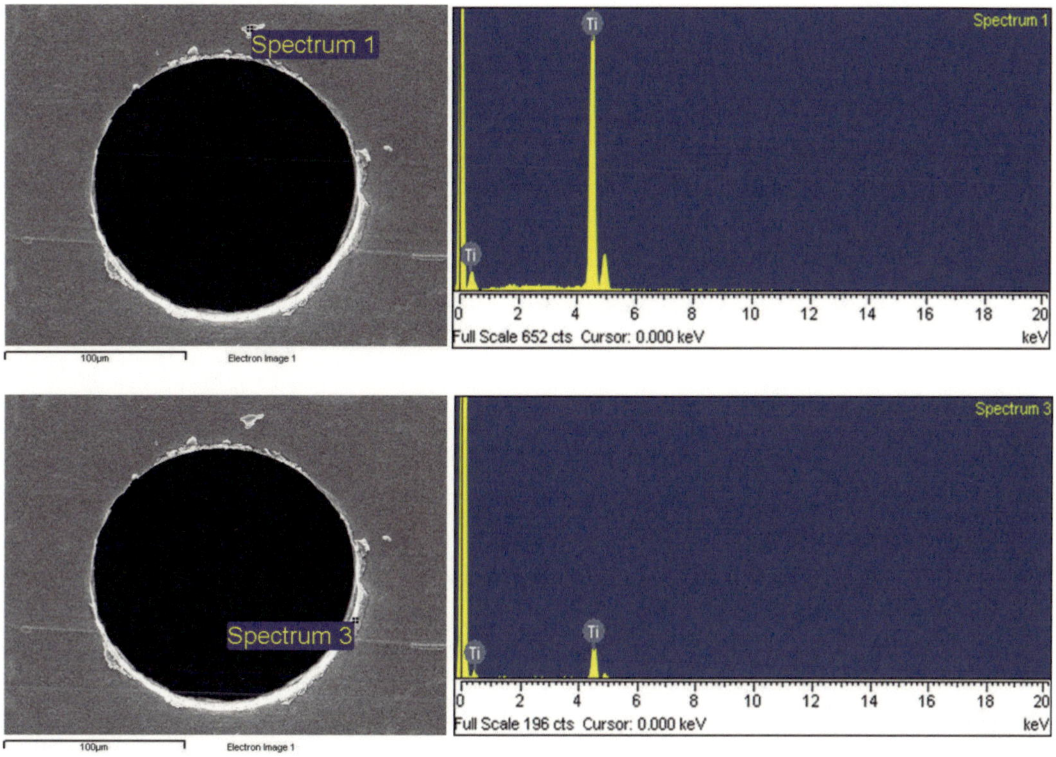

Figure 8. EDS results of debris in the back side.

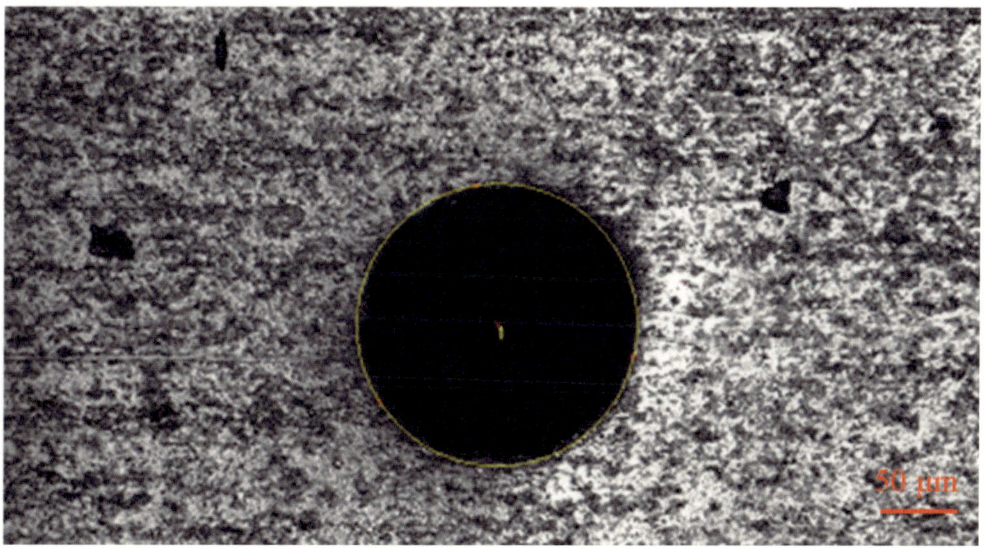

Figure 9. Profile of microhole punched by 75WC/25Co micropunch measured by OLS3000.

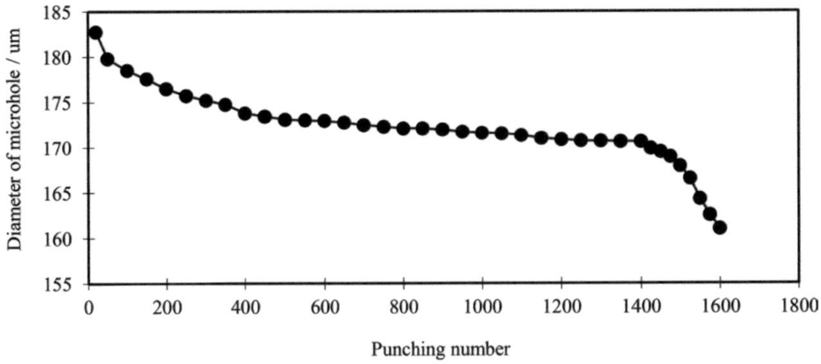

Figure 10. Relationship between diameter of microhole and punching number.

4. Conclusion

The effect of cobalt on the morphology of microhole formed by punching had been researched with three types of WC/Co micropunches. It shows that the quality of micropunch with 75% vol. WC + 25% vol. Co is distinctively better than the other two types. Meanwhile, it reveals that due to the optimal joint-contribution of WC and Co, the morphologies of the microholes punched by 75WC/25Co micropunches not only satisfy with the practical requirements but also the relevant punch using life is at the highest level among the three types of micropunches. As for 75WC/25Co micropunch, when the punching number exceeds 1525, the serious wear of micropunch occurs with the wear loss of Co and WC. Moreover, with the punching numbers further increment, the dominant factors of the wear loss would mainly rely on the easily peeled off WC due to the serious loss of Co.

Acknowledgements

The work is supported by the Strategic Research Grant (SRG) from City University of Hong Kong (Grant no.: 7004598).

Author details

Kelvii Wei Guo* and Hon Yuen Tam

*Address all correspondence to: kelviiguo@yahoo.com

Department of Mechanical and Biomedical Engineering, City University of Hong Kong, Hong Kong

References

[1] Mark JJ. Microfabrication and Nanomanufacturing. Boca Raton FL: CRC/Taylor & Francis; 2006

[2] Christophe G. Microsystems Engineering: Metrology and Inspection III. Bellingham Wash: SPIE; 2003

[3] Madou MJ. MolDock: Microfabrication challenge. Analytical and Bioanalytical Chemistry. 2004;**378**:11-14

[4] Joo BY, Rhim SH, Oh SI. Micro-hole fabrication by mechanical punching process. Journal of Materials Processing Technology. 2005;**170**(3):593-601

[5] Brousseau EB, Dimov SS, Pham DT. Some recent advances in multi-material micro- and nano-manufacturing. International Journal of Advanced Manufacturing Technology. 2010;**47**:161-180

[6] Fonda P, Katahira K, Kobayashi Y, Yamazaki K. WEDM condition parameter optimization for PCD microtool geometry fabrication process and quality improvement. International Journal of Advanced Manufacturing Technology. 2012;**63**:1011-1019

[7] Guo W, Tam HY. Effects of extended punching on wear of the WC/Co micropunch and the punched microholes. International Journal of Advanced Manufacturing Technology. 2011;**59**:955-960. DOI: 10.1007/s00170-011-3567-0

Chapter 7

A Review of the Radionuclide, Cobalt-60, as a Fine-Sediment Tracer

Philip Greenwood

Additional information is available at the end of the chapter

http://dx.doi.org/10.5772/intechopen.71304

Abstract

The chapter firstly outlines the global crisis concerning accelerated soil erosion, the implications that this could have on longer-term food security and ways that soil loss has been quantified originally through catchment-scale monitoring but with a recent shift toward the use of sediment tracers. It then briefly reviews some of the most commonly used tracers in sedimentation studies, before focusing on artificial, gamma-emitting radionuclides, and in particular, on Cobalt-60 (Co-60). Some historical background information on previous uses of Co-60 are then provided, and the suite of key environmental characteristics that, from the perspective of studies in hydrology and geomorphology, make Co-60 a potentially attractive candidate for fine-sediment tracing. The chapter then outlines and reviews three ways in which Co-60 has been experimentally applied in varying erosion and sedimentation scenarios where most of the more commonly used tracers would be unsuitable or would lack the level of sensitivity needed to return meaningful data. It then outlines some of the notable drawbacks associated with using Co-60, before highlighting refinements and prospects for future work. The chapter finally concludes by evaluating the versatility and efficacy of Co-60 as a fine-sediment tracer.

Keywords: soil erosion, land degradation, Cobalt-60, fine-sediment tracer, sediment transfer mechanism, river floodplain, remobilisation, earthworm casts, livestock-poaching, sediment-source

1. Introduction

1.1. The importance of sediment studies

The mobilisation, transportation and deposition of fine sediment (defined here as particles and aggregates <2 mm dia.) by rainfall and surface run-off are natural processes but are often accelerated by certain human activities that include *inter alia*, deforestation, modern agricultural practises and/or adopting inappropriate land uses [1, 2]. The recognition that soil and sediment both represent a vital resource has led to concerns that accelerated soil erosion, along with the concomitant degradation of agricultural land, poses significant threats to sustainable food production to long-term global food security [3, 4] and to the quality of the environment *per se* [5]. Although inherently difficult to quantitatively express the degree to which land is, or has been, degraded at the global scale, varying authors have offered numerous statistics in an attempt to judge the severity of the problem. In the mid-1980s, for instance, Brown [6] concluded that global soil depletion annually exceeded the amount that could be replenished under natural conditions by around 23 billion tonnes each year. In the 1990s, Oldeman et al. [91] debated that two billion hectares (ha), representing 22.5% of all productive land on Earth, was more degraded than it was in the mid-1940s. More recently, Imeson [7] reported that global agricultural cropland productivity is diminishing by an average of 5% each year, principally by water-driven soil erosion but undoubtedly exacerbated by modern agricultural practises, whereas others such as Cherlet and Ivits [8] concluded that approximately 24 billion tonnes of fertile soil is lost to erosion and degradation each year, which, they speculate, could lead to a 12% reduction in global food production in the next 25 years. Against an expanding world population reliant on finite land on which to produce food and offer continuing food security [9], it is now acknowledged that the accelerated loss of a non-renewable resource such as soil is unsustainable which not only leads to land degradation and reduced food production *per se*, but also can lead to many environmental problems seemingly unconnected to agriculture and food production [10].

Reasons for many environmental problems can be attributed to fine to fine sediment which acts as a vector for the transfer and off-site conveyance of nutrients, which, when in concentrated form, can readily act as a pollutant. The loss of nutrient-rich fine sediment from agricultural land not only leads to a gradual reduction in the fertility and yield of crops from that land, but also the transfer of fine sediment carried away in agricultural run-off inevitably contributes to reduced water quality in receiving watercourses and to the widespread degradation and eutrophication of aquatic habitats downstream of its entry point [11]. Indeed, excessive concentrations of fine sediment, carried in suspension from field to watercourse, also have the ability to physically disrupt the internal functioning of aquatic environments and can act as a diffuse source of contamination in both terrestrial and aquatic ecosystems alike [12].

In response to the threat that excessive loss of soil poses to the environment, hill slope and catchment processes have come under increasing scrutiny in their role as sediment sources and sinks. Against the need to improve water quality in fluvial and aquatic environments, legislation such as the European Union (EU) Water Framework Directive (WFD) [13] has sought to promote a holistic approach to river basin management, with the net effect of reducing pollution and the amount of sediment entering into watercourses, in order to preserve and enhance their biodiversity and amenity status and sustain their future well-being for the benefit of all stakeholders [13]. Motivated by this legislation, the formulation of catchment-scale management plans has highlighted an urgent need for effective sediment control strategies to manage and mitigate the deleterious effects of fine sediment in the wider environment but particularly to reduce agricultural run-off from leaving the field (**Figure 1**) and ultimately entering aquatic environments. The effectiveness of such measures, however, ultimately depends on the research community being able to identify and thereafter quantify the spatial and temporal flux of fine sediment within catchments [14].

Prognoses for estimating sediment movement within catchments have typically relied on monitoring techniques, with the focus on quantifying sediment yields at catchment outlets [15]. Although this approach has provided an initial framework around which overall annual sediment budgets have been constructed, recognition that not all mobilised fine sediments actually reaches the river but, instead, may be temporarily stored on depositional areas that have directed increasing attention to the internal functioning of river catchments themselves [11]. Researchers have thus sought to gain a more detailed understanding of the myriad sediment conveyance mechanisms that link upstream mobilisation of fine sediment to the downstream sediment yield and to determine the relative efficiency of such mechanisms in conveying sediment from the hill slope to the river [16]. However, identifying and elucidating how important those links are remains one of the most complex and least understood components of a catchment system. Importantly, it has also highlighted limitations in the ability of many existing monitoring techniques to provide sufficiently detailed information to enable more meaningful sediment budgets that integrate sediment sources and sinks, and the conveyance pathways interposed between them to be constructed. Those limitations have necessarily prompted a shift from purely monitoring the movement of fine-sediment toward tracing its movement, but the latter approach was very much viewed as mutually complementing the former approach rather than replacing it [17, 18].

Figure 1. Agricultural run-off leaving a recently cultivated field is often laden with fine-sediment and sediment-bound nutrients, such as phosphorous and nitrogen.

2. Sediment tracers

A tracer can be defined as any physical object or chemical substance capable of being introduced into a physical, chemical, or biological system, in order that it may be subsequently tracked over a given timeframe and provide information that can be inferred by its redistribution or dispersal [19, 20]. However, the suitability of any tracer depends on its ability to meet certain requirements. From the perspective of work presented and reviewed here, arguably the most fundamental characteristics are:

i. ease of identification;

ii. the ability to mimic the behaviour of the media being traced as closely as possible, yet remain distinguishable from its surroundings; and

iii. an ability to remain as conservative in the environment as possible [19–21].

A wide variety of materials and substances meet those requirements, and the choice of tracer will ultimately depend on a number of things. Arguably, those that are most fundamental are the cost and availability of the tracers themselves, the area under investigation, and hence, the area over which the tracers can be applied and recovered, but also the textural characteristics of the sediment being traced and whether it is fine, medium, or coarse grained. Since most sediment-related environmental problems relate to the fine-sediment fraction, a very brief review follows that accordingly focuses on some of the most commonly used types of tracers that have been used to study the movement of fine sediment.

2.1. Passive and active tracers: a brief review

For clarity, sediment tracers can be divided into two sub-groups: those that are 'passive' and those that are 'active'. Definitions of passive and active tracers can vary, depending on the research discipline making the distinction, the nature of the research being undertaken, and the frequency and stage during an investigation at which measurements of the movement or dispersal of the tracer/sediment can be undertaken (i.e., during or after an event). In light of this uncertainty, the following text provides a working yet extended definition of passive and active tracers, in the context of the work presented and reviewed in this chapter.

2.1.1. Passive tracers

Passive tracers are defined as those that have no additional energy source external to the combined force moving both sediment and tracer together. In essence, therefore, passive tracers are inert, and their redetection cannot be undertaken via an internal function of the tracer itself. Passive tracers come in many forms, but ideally their physical size and density should match, as closely as possible, the texture of the soil being investigated. Examples of some of the more commonly used passive tracers in soil erosion studies include metal cubes (e.g. [22]), plastic beads (e.g. [23]) and coloured gravel (e.g. [24]). As such, passive tracers are often commonly referred to as physical point tracers, and importantly, their inert nature means that methods of recovery and translocation distance typically involve relocation by physical-based approaches. For fine, particulate-sized material, this may involve systematically removing small quantities of surface material at different locations along a line of erosion and then recovering the tracer material from the mobilised/deposited sediment. Rates of tracer displacement, and hence rates of mobilisation, can then be inferred based on changes in the concentration of the tracer relative to its original concentration prior to the erosion event. For larger physical point tracers, their relocation may involve more elementary approaches, such as field-walking, to observe and then record the location of individual tracers relative to their original position before the erosion event. By necessity, however, all recovery methods are typically performed after one or a series of consecutive erosion events. As the recovery of passive tracers is exploratory, it is invariably also invasive for accessing and trampling the site, and disturbing and removing at least some of the host sediment. Consequently, this typically restricts the use of passive tracers to measuring erosion over single, yet high energy, events, or over events that occur in quick succession. For this reason, passive tracers have proven to be ideally suited for measuring rates of soil translocation during tillage operations on hill slopes (e.g. [25]).

Substances known as rare earth elements (REEs) have proved particularly useful for studying erosion over relatively short timescales, particularly in areas where high erosion rates reduce the effectiveness of other techniques (e.g. [26, 27]). Numerous characteristics have contributed to their popularity, and these include their availability, their conservative behaviour, and their ability to bind to individual clay particles but also to coalesce and thereafter mimic the behaviour of aggregates. Low background concentrations in most environments also permit high analytical sensitivity [28]. Other positive attributes include their inert nature, benign characteristics, and the ability to detect different REEs simultaneously, usually by inductively coupled plasma-mass spectrometry (ICP-MS). Disadvantages with this type of material, however, include the need for detailed sampling campaigns covering the full extent of an area under investigation to provide a sufficient level of spatial resolution. As this represents a prerequisite, to ensure representative erosion estimates are obtained from across the entire study area and not just erosion 'hot-spots', this usually limits the area over which REEs can feasibly be applied to just a few 10 s of m^2. In addition to this, larger areas are difficult to access to physically take samples for analysis without altering or manipulating the very soil surface on which subsequent data may be based.

2.1.2. *Active tracers*

In contrast to passive tracers, active tracers are defined as any material or substance whose redetection and subsequent measurement are facilitated by an internal function of the tracer itself [19]. Such internal functions often have the ability to operate both independently and irrespective of environmental conditions [29]. This permits modes of identification and measurement to be conducted using a variety of different techniques, many of which are non-invasive and/or non-destructive and can thus be performed repeatedly, either in-situ and/or remotely. Those attributes permit measurements to be undertaken at different stages during the operational phase of an experiment; an attribute that affords researchers the opportunity to iteratively monitor the movement or dispersal of the tracer over a wider range of timescales. In most instances, therefore, the main characteristics evidenced by active tracers are often viewed positively, as they confer a major advantage over passive tracers, by allowing repeat measurements to be performed over one or a sequence of erosion events occurring in quick succession or even over slightly longer (i.e., seasonal) timescales [30, 31].

Some examples of active tracers previously used in erosion experiments include magnetic powder (i.e., magnetite) sorbed onto small polystyrene beads and incorporated into a host soil at known concentrations. Changes in the magnetic susceptibility of the soil relative to the original value thereafter provide a means of determining the quantity of mobilised soil (e.g. [32]). Similarly, crushed magnetite with textural characteristics similar to a host soil has been introduced at a point-source location at known concentrations within a bounded erosion plot. Samples of run-off-derived sediment were recovered from the plot outlet after a series of erosion events and re-measured. An elevated magnetic signal relative to the background level from the host soil allows researchers to not only quantify the travel distance of the labelled material but also infer how much of the magnetised material has been mobilised (e.g. [33]).

In addition to the above, ceramic prills have been labelled with dysprosium oxide (Dy_2O_3), which, after exposure to rainfall, have been recovered and subjected to neutron activation analysis (NAA), in order to make the dysprosium oxide (temporarily) radioactive, thereby allowing soil redistribution rates to be determined (e.g. [34]). Alternatively, a more complex labelling approach has been undertaken, whereby Europium (Eu_2O_3) was incorporated into molten glass, which was then cooled and grounded to different sizes to represent soil aggregates. Prior to cooling, lead was also introduced into the mixture, which facilitated adjusting the bulk density of the tracer to the host soil [35]. Similar to the previous approach, the mixture was then subjected to NAA; transforming the Europium into the gamma-emitting radionuclide, Eu-125 m (^{125m}Eu), whereupon the labelled material was used in a variety of erosion investigations.

2.1.3. Fallout radionuclides as sediment tracers

A number of globally distributed (i.e. fallout) radionuclides, namely lead-210 (^{210}Pb), beryllium-7 (^{7}Be), and caesium-137 (^{137}Cs), naturally decay by γ-radiation and express a very high affinity for fine sediment. Consequently, all three sorb to individual clay and silt particles, and aggregates, rapidly and very strongly on contact. This means that in undisturbed soils, most of the fallout inventory remains within the upper part of the soil profile. Both attributes render all three radionuclides useful for studying the redistribution of surface sediment. Over the last ca. five decades, therefore, researchers have increasingly sought to harness those characteristics through developing and refining ways in which those particular fallout radionuclides can be exploited and used to investigate soil redistribution processes at the field to catchment scale (e.g. [36, 92]). Continual refinements to the original technique have steadily provided researchers with an increasingly deeper understanding of sediment dynamics operating both within and between river catchments. Reasons for many of these process-based advances have largely been attributed to the use of tracers that undergo nuclear decay. This produces a signature along parts of the electromagnetic spectrum (EMS) that is not only unique and provides means of identifying each radionuclide, but the rate of decay is specific to each radionuclide and operates independently of fundamental environmental conditions; namely temperature, pressure, light, and moisture, which can unduly alter or influence the behaviour of other tracers, while investigations are still in progress [29]. As such, the rate at which certain radionuclides undergo nuclear decay has duly allowed a temporal perspective to be established for many environmental processes [37]. Indeed, timeframes have now been attributed to a wide range of sedimentation-related mechanisms once considered too slow and hence immeasurable or simply too innocuous to merit closer attention [11, 38–42]. In the case of the anthropogenic ^{137}Cs ($t_{0.5}$ = 30.2 years) and the lithogenic ^{210}Pb ($t_{0.5}$ = 22 years), those two medium-term fallout radionuclides have permitted field and catchment-scale information on sediment redistribution to be estimated retrospectively over timescales of ca. 50 and 100 years, respectively (e.g. [43, 44]). This has provided researchers with means of quantifying a number of fundamental catchment processes that include estimating the volume of sediment deposited across river floodplains during flood events, quantifying the residence time of deposited sediment on river floodplains and seeking evidence of catchment responses to longer term land-use changes [44]. However, while those factors represent major advances in our understanding of the way in which catchments function, estimates of erosion and deposition using fallout ^{137}Cs and ^{210}Pb must, by necessity, be presented as yearly averages over the aforementioned time periods only. Consequently, those approaches are unable to quantify soil redistribution over short or event-based timescales, or where erosion may have occurred due to sudden changes in land-use conditions [45]. In situations where event-based redistribution data are required at the field-scale and larger, recent developments using the short-lived cosmogenic fallout radionuclide, ^{7}Be ($t_{0.5}$ = 53 days), have proven successful (e.g. [46, 47]); to the point where it is now established as an additional technique that compliments the use of the two medium-term tracers described above. However, a major disadvantage associated with fallout ^{7}Be is that it can be readily sequestered by surface vegetation [48]. Depending on the type and density of the vegetation at a given study area, its presence can adversely influence the amount of inventory that actually reaches the soil surface. Since this can initiate changes in the depth distribution of ^{7}Be in the soil profile, while the investigation is still in progress [49], small-scale variations in inventory values may produce erroneous results. Consequently, studies using ^{7}Be have tended to be focused on areas where soils have recently been cultivated and the surface is still free of vegetation, or areas where vegetation cover is sparse or crop residue is low [48].

2.1.4. *Artificial gamma emitting radionuclides as sediment tracers*

The use of artificial gamma-emitting radionuclides as sediment tracers, or to clarify, radionuclides that have been deliberately introduced into or across a small predefined study area or used to 'label' small quantities of fine sediment with low-level radiation have the potential to overcome many of the limitations associated with some of the more commonly used tracers outlined and reviewed above. They are not, however, designed to replace existing tracers but instead should be viewed an additional tool in the proverbial 'toolbox' of tracers and tracing techniques that researchers can draw upon during instances where detailed information about sediment redistribution is required at the small scale and at high temporal resolution.

There are a number of notable advantages associated with certain artificial radionuclides as fine-sediment tracers; particularly, species that undergo nuclear decay by emitting gamma radiation. Among the most noteworthy advantages included in their limited distribution in the wider environment, the level of convenience afforded by their use in soil erosion investigations and, above all, the conservative behaviour, and hence, the ability of certain artificial radionuclides to sorb to individual soil particles and aggregates rapidly and strongly. This attribute enables them to faithfully mimic the behaviour of fine sediment during an erosion event without physically perturbing or adversely influencing its behaviour while in transit. In this respect, sediment labelled with a suitable artificial radionuclide, for the purpose of tracing its movement, arguably represents one of the most effective ways of precisely matching the tracer characteristics to the sediment being traced [38]. This particular advantage can be attributed to the 'minimal mass theory' proposed by authors such as Sauzay [20], Courtois (1973) and echoed latterly by McCubbin and Leonard [50]. The general principle has previously been alluded to in an earlier section, but to explain in more detail, it argues that the behaviour of any object or substance used as a tracer must be as compatible as possible with the medium being traced in order to derive accurate and representative information. Because individual gamma photons carry very little mass but emit a powerful and unique signature that is distinct to each radionuclide, any labelled sediment that is subsequently redistributed will carry vestiges of that same signature, albeit not as strongly as the original signature before dispersal. By measuring and comparing the strength of the signature (also referred to as an 'inventory') before and after an erosion event, the change not only indicates that labelled material had been dispersed but can also be used to quantify the mass of sediment dispersal [30, 31, 38, 50, 51]. The process of nuclear decay is particularly useful to tracer scientists undertaking studies of certain sedimentation processes. This is because one of the most commonly occurring, ubiquitous and hence important erosion mechanisms is particularly subtle, since it occurs on shallow slopes under low-energy conditions. The process, which is known as 'sheetwash' or 'inter-rill erosion', preferentially entrains small particles that have previously been detached by raindrops as they impact against the soil surface. It then carries the liberated sediment under the influence of gravity downslope in a thin layer of water (often <3 mm deep) [52]. Sheetwash commonly occurs on all low-gradient slopes and is the precursor to the more visually obvious process known as rill erosion, where topography serves to concentrate the overland flow, whereupon it becomes much faster flowing and has sufficient hydraulic power to incise its way down through the soil profile [38], indiscriminately detaching and transporting fine-sediment of all sizes (**Figure 2**). In areas where deposition of sheetwash-derived sediment takes place, gamma-emitting radionuclides can be particularly useful for measuring certain sedimentation processes. This is because only a few gamma photons are required to identify the species of radionuclide. Arguably, therefore, one of the biggest attributes associated with radionuclides that emit γ-radiation is the relative ease with which gamma photons can be detected, measured, and used to speciate individual radionuclides. Due to the fact that gamma photons are ejected at very high energy but have negligible mass, they are highly penetrative and easily travel through most substances, including thin layers of deposited sediment. Consequently, re-identification of small quantities of pre-labelled sediment is possible, even in depositional areas, where subsequent burial by unlabelled material may have occurred [39, 40]. Because of the above attributes, soil that has been pre-labelled, even with relatively low levels of a suitable radionuclide, will theoretically not only behave in a manner that is identical to unlabelled soil, but also its redistribution or burial can be detected and accurately measured with a level of sensitivity that far exceeds the majority of other tracers.

Figure 2. When un-concentrated overland flow becomes concentrated, it forms a network of small channels called rills which gradually incise into the soil profile and transports the eroded material downslope.

Despite numerous demonstrations of the versatility and effectiveness of artificial radionuclides to provide reliable information on fine-sediment redistribution over short timescales, however (e.g. [30, 31, 38, 40, 53–56]), their use in erosion investigations has attracted surprisingly little attention over the last ca. 50 years. Reasons for the slow uptake remain unclear but may be attributed to their controversial nature and the commensurately stringent suite of laws that govern the purchase and subsequent release of open-source radionuclides in the environment (e.g. [57]), difficulties associated with uniformly labelling soils and sediments prior to undertaking in- or ex-situ erosion experiments and the limited areal extent over which they can realistically be applied (*cf.* [54]). With the last constraint in mind, research using artificial radionuclides has tended to focus on small-scale investigations, where the area of application may be less than one square metre [58] and certainly no larger than a few hundred square metres [54]. Despite this apparent limitation, numerous workers have circumvented this scale constraint by utilising a point-source tracing approach, whereby the radionuclide may be applied to a network of small areas across, for instance, a field-scale study area (e.g. [38, 40]). Such an approach has proven to be both convenient and very effective at iteratively documenting soil redistribution data over one or a more consecutive rainfall or flood events, particularly if care is taken to select representative locations within the larger study area.

2.1.5. *The anthropogenic radionuclide, Cobalt-60*

Cobalt-60 is an anthropogenic, gamma-emitting radionuclide with a relatively short half-life ($t_{0.5}$) of approximately 5.26 years [59]. It is classified as an activation product that is inadvertently manufactured in nuclear reactors by neutron activation of the naturally occurring isotope, Cobalt-59 (^{59}Co). Cobalt-59 is a stable metal element that is relatively ubiquitous in the environment and occurs in various minerals and metal ores and typically remains as an impurity within most finished metal products. Metals containing ^{59}Co are frequently, and inadvertently, used in the construction of nuclear reactors. Upon exposure to the intense radiation field within the reactor, ^{59}Co absorbs a free neutron into its nucleus, transforming it into the unstable radioisotope, ^{60}Co. Once transformed, ^{60}Co begins to emit ionising radiation in the form of beta particles (β^-) and gamma radiation (γ-radiation). After 5.26 years has elapsed (i.e. one half-life), radioactive decay transforms 50% of the original ^{60}Co nuclei into its stable progeny, nickel-60 (^{60}Ni) [60]. Cobalt-60 emits γ-radiation at photo-peaks of approximately 1173 and 1332 kilo electron volts (keV). The energy field, which is defined as the proportion of ^{60}Co atoms produced per 100 disintegrations, is ca. 100% at each photo peak. It is therefore theoretically possible to obtain accurate results from samples emitting activities as low as 1.0 Becquerel (Bq) kg^{-1} of (dry) sediment, assuming long (≥1 day) analytical counting times on a laboratory-grade gamma spectrometer.

With regards to its presence in the environment, small quantities of ^{60}Co are discharged under licence as effluent into marine environments from nuclear reprocessing plants such as La Hague, in northern France, and Sellafield in Cumbria, UK [61–63]. Discharges of ^{60}Co can be detected in close proximity to its release point in samples of seawater [61, 64], in surface sediments or from shallow sediment cores [62, 63]. Some researchers have exploited these licenced releases and used the signature of ^{60}Co as a tracer or geochemical marker to elucidate varying environmental processes [61, 62, 64, 65]. Further to those licenced releases noted above, areas in close proximity to disused underwater nuclear test sites also represent additional, yet highly localised, sources of ^{60}Co [63].

Owing to its purely anthropogenic origin, its relatively short half-life and limited discharge sources, ^{60}Co does not, therefore, naturally occur in the wider environment. This not only simplifies things when undertaking sediment tracing investigations, but it also represents a notable advantage for a number of other reasons. Principally, an absence of background contamination removes the possibility of obtaining spurious data due to spatially variable background levels (*cf.* [66]). This permits both tracer inputs and outputs to be very precisely quantified [20]. This also facilitates the likelihood that subtle changes in radionuclide inventories after an erosion event could signify very subtle movement of labelled sediment, instead of being potentially attributable to spatial variations in background concentrations. Further to the above, an absence of background contamination also permits the use of lower initial dose rates (i.e. the amount of inventory sorbed to a given mass of sediment, or sorbed over a given area) when undertaking tracing investigations [53]. Importantly, all the above factors collectively contribute to potentially increasing the analytical sensitivity, to improving data accuracy, and increasing the longevity of the monitoring campaign.

A desktop review of ^{60}Co reveals that a substantial amount of literature has been compiled which documents the characteristics, behaviour, and mobility of ^{60}Co under varying environmental scenarios (e.g. [59, 63, 67–69]). Most relevant to work presented in this chapter are the rate and strength of retention (sorption) to fine sediment, the solubility and migration of ^{60}Co, and its rate of biological uptake (bio-uptake) and accumulation, and thus its potential toxicity to organisms.

With respect to the first two parameters, the solid/liquid distribution coefficient (K_d) values have been used to describe the mobility, sorption, and desorption of ^{60}Co in response to a range of environmental and physical controls. These range from particle size characteristics of the material being labelled and traced, the organic matter (OM) content, pH, Eh, temperature, moisture and time [59, 68, 70]. K_d values for ^{60}Co, obtained from wide ranging soil-types, indicate that sorption is typically high to very high [59, 69]. Numerous researchers have reported that pH represents a major controlling factor that can frequently dominate the rate at which sorption occurs, especially in acidic (i.e. ≤pH 6) environments [59, 68]. However, the influence exerted by pH diminishes very rapidly from ca. pH neutral and upwards, and by ~pH 8, sorption to sediment is reportedly ~100% [59, 71, 72]. Under environmental conditions typically encountered in agricultural environments, therefore, sorption of ^{60}Co to soil is both rapid and considered to be largely irreversible [59, 69, 71, 73]. The environmental characteristics of ^{60}Co have resulted in it being used as a tracer in a wide variety of investigations over approximately the last six decades. To demonstrate this, some investigations from the available literature have been compiled and are listed in **Table 1** in the order of publication.

Although the list is neither exhaustive nor comprehensive, it provides some indication of the diverse applications in which ^{60}Co has previously been used and thus highlights its potential versatility as a tracer of many environmental processes. Owing to the relatively limited number of ^{60}Co sources and pathways for its entry into the environment, however, very little literature exists regarding its biological uptake, accumulation and potential toxicity to organisms. Studies that have made reference to those particular characteristics do not discuss its toxicity potential in great detail (e.g. [51, 55, 56, 69, 74, 75]) but infer that its likely rate of transfer and accumulation from soils into and through the food chain and its injurious potential to organisms at low concentrations are minimal. This inference is corroborated, albeit by logic alone, via the wide variety of ecological and medical tracing applications where ^{60}Co has been introduced onto, or into, numerous organisms (including humans) during tracing investigations, some of which are listed in **Table 1**.

With regard to the use of ^{60}Co in sediment tracing studies, Toth and Alderfer [55] document two of the earliest known applications, with the first publications providing details of the development of a procedure for uniformly tagging (labelling) water stable aggregates (WSAs) by immersion into solutions of ^{60}Co and water. The success of the labelling technique culminated in the same authors conducting a year-long investigation into the formation and breakdown of the ^{60}Co-labelled WSAs [56]. The reported success of both the labelling technique and the tracing application provided an initial indication of the potential value of ^{60}Co as a candidate for sediment tracing work of a similar nature. However, the apparent dearth of sediment-related tracing investigations since those published in 1959 implies that the versatility and possible value of ^{60}Co may have been ignored or inadvertently overlooked by subsequent researchers.

The following section now briefly describes procedures for sample processing, as well as, arguably, the most accurate and cost-effective means of measuring low level radioactivity in soils and sediments, namely by gamma spectrometry.

Data source	Publication year	Nature of investigation
Arnason et al. (in Ref. [51])	1950	Determine the movement of wireworms (*Ctenicera* spp.) in soil
Rings and Layne (in Ref. [51])	1953	Determine dispersal of beetles (*Conotrachelus* spp.) in woodland
Sullivan (in Ref. [51])	1953	Determine dispersal and habitat range of beetles (*Pissodes* spp.) in woodland
Babers et al…	1954	Determine dispersal of Boll Weevils (*Anthonomus* spp.)
Green and Spinks (in Ref. [51])	1955	Determine the movement of wireworms (*Agroites* spp.) in soil
Green et al. (in Ref. [51])	1957	Determine the movement of moth larvae (*Rhyacionia* spp.) in soil
Green et al. (in Ref. [51])	1957	Determine the dispersal of mature moths (*Rhyacionia* spp.)
Toth and Alderfer	1959a	Establish a procedure for labelling water stable aggregates by immersion
Toth and Alderfer	1959b	Monitor the formation and breakdown of labelled water stable aggregates
Traniello et al. (in Ref. [51])	1985	Determine territorial feeding patterns of termites (*Reticulitermes* spp.)
Rosengaus et al. (in Ref. [51])	1986	Determine of termites (*Reticulitermes* spp.)
Thorén et al. [94]	1991	Determine location of cancerous brain tumours in humans
Cundy and Croudace	1996	Wastewater effluent as a geochemical marker for estimating sedimentation rates
Alam et al. [90]	2001	Determine rates of nutrient uptake in agro-crops
Capowiez et al.	2001	Determine 3-dimensional trajectories of burrowing earthworms
Sattar et al.	2002	Determine feeding habits of termites (*Heterotermes, Microtermes* and *Reticulitermes* spp.)
Thompson et al.	2002	Wastewater effluent as a geochemical marker for estimating sedimentation rates
Greenwood et al.	2008	Used as a fine-sediment tracer
Schreiner et al. [93]	2009	Tomotherapy cancer treatment in humans
Greenwood	2012	Used as a fine-sediment tracer
Greenwood et al.	2013	Used as a fine-sediment tracer
Greenwood et al.	Accepted	Used as a fine-sediment tracer

Table 1. Selected studies, published from 1950 to the present day, that have used the artificial radionuclide, ^{60}Co, as a tracer of biological or environmental processes.

2.2. Sample processing & methods and units of measurement

As previously mentioned, each gamma-emitting radionuclide produces one or more photo peaks, or signatures, at known locations along the EMS, which allows each radionuclide species to be readily identified and their prevalence to be quantitatively determined. There are several methods of determining low-level radioactivity from soil and sediment samples. One of the most accurate and cost-effective methods is by gamma spectrometry using detectors fitted with a high-purity germanium crystal (HPGe) [76]. Sample analysis is a relatively simple procedure that involves minimal prior sample preparation. It is also non-destructive, which allows repeat measurements to be undertaken on individual samples. For laboratory-based analysis, the key stages involve drying, gently disaggregating and screening each sample through a 2-mm diameter sieve. Sieved material should be weighed to a predetermined quantity (i.e. ca. 100 g) and to an accuracy of one decimal place [77]. For reasons of precision, samples presented to a detector should be of similar mass (i.e. within ±0.1 g), and radiometric assays should be performed in containers of identical proportions, so that the sample geometry, and hence, the distance from the sample to the germanium crystal, remains constant. Failure to comply with this rule can profoundly influence the precision of measurements, leading to erroneous results [76] The sample is then placed in very close proximity to the germanium crystal, which is housed in the detector head (**Figure 3A**). Gamma-photons ejected

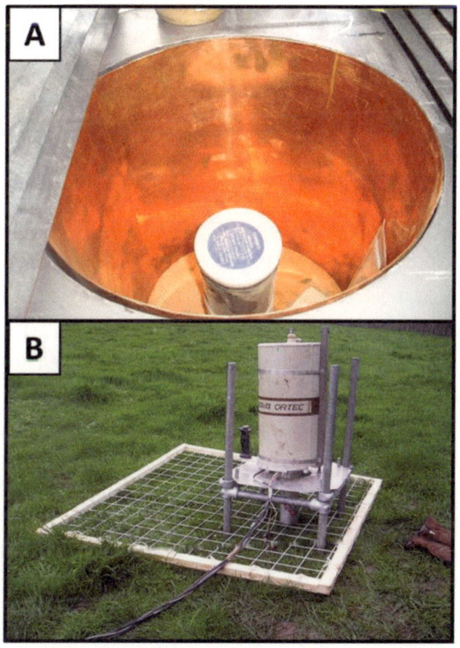

Figure 3. (A) The inside of a laboratory-grade gamma-spectrometer showing the top of the germanium crystal onto which a sediment sample is placed for analysis, and the lead and copper surround, with sliding cover lids to reduce interference from incoming background gamma radiation. (B) An in-situ gamma spectrometer is placed with the detector head facing downwards and close to the soil surface. Establishing and maintaining a consistent measurement protocol throughout a monitoring campaign is necessary to ensure that results are reliable.

from the sample interact with the electrically charged germanium crystal, which in-turn emits signals or pulses that correspond to the energy of the incoming gamma photons. These are routed through a preamplifier to an amplifier and then to a multi-channel analyser (MCA), whereupon the signals are displayed as a spectrum on a computer screen. Since the energy of the emitted photons is unique to each radionuclide, those that are present in the sample can be identified and their respective activity quantified. Field-based areal activity measurements can be performed in-situ using a field-based gamma spectrometer, shown in **Figure 3B**. Although no sample preparation is required when taking in-situ measurements, the crucial factor to obtain reliable results is to ensure that the radiometric protocol adopted at the beginning of the investigation remains consistent throughout the monitoring campaign (*cf.* [16, 40, 45]). More information about in-situ measurement protocols is given in the following section and can also be found in Greenwood [16] and Greenwood et al. [40].

The most convenient and appropriate measurement unit, which is also recognised by the International System of Units (SI), for assessing low-level radioactivity emitted by any radionuclide in soils and sediments is the Becquerel (Bq). One Bq is equal to one decay per second (dps). Because this unit is small, it is often convenient to use additional prefixes, such as the kBq (i.e. 10^3 Bq) or MBq (10^5 Bq). It is also often necessary to use an additional prefix, for instance, where the concentration of fallout radionuclides is being determined from small soil samples at typical environmental activity levels, which enable the activity concentration to be measured and expressed in milli-Becquerels (mBq) that are equal to 10^{-3} Bq.

3. Practical applications

The following sections now describes and reviews examples of three field-/laboratory-based prototype tracing techniques using ^{60}Co in order to demonstrate the novel and diverse ways in which it has recently been used as a fine-sediment tracer. The underlying motivation for each investigation was to acquire essentially unique information that would otherwise be unobtainable if cruder or less-sensitive types of tracer were used. As indicated earlier, the attribute of sensitivity has proven to be particularly beneficial during attempts to study, for instance, the movement, redistribution or remobilisation of fine sediment in environments where the energy regime may be low, and hence, redistribution rates are likely to be commensurately small, or where mobilisation occurs over very short (i.e. event based) timeframes.

3.1. Example 1: assessing flood-derived sediment remobilisation on river floodplains

3.1.1. Experimental aim

The aim of this investigation was to explore the potential for documenting the extent to which fine sediment, recently deposited on a river floodplain during an overbank flood, might be subject to remobilisation and redistribution. Further details regarding the approach and the findings can be found in Greenwood et al. [39, 40]. Motivation for the investigation arose due to the realisation that river floodplains not only represent important stores of fine sediment but also act as stores of sediment-associated nutrients and contaminants [78]. Consequently, increasing awareness of their environmental importance has resulted in attempts to quantify the capacity of river floodplains to store fine sediment and to determine the residence time of the deposited material [79]. While those factors have been satisfactorily quantified to a certain extent using the aforementioned medium-term fallout radionuclides, ^{137}Cs and ^{210}Pb, measuring sediment accretion rates over shorter timescales and assessing the extent to which recently deposited sediment may be remobilised during subsequent flood events have, by contrast, received limited attention. Reasons for the oversight have traditionally been attributed to difficulties in tracing the movement of small quantities of fine sediment over short time periods. Scope therefore existed to design an experiment for the purpose of testing ^{60}Co on this hitherto unrecognised and unstudied sedimentation scenario by undertaking the following approach.

3.1.2. Method

The approach involved pre-labelling small (i.e. ca. 200 g) quantities of fine sediment with 80 Bq of ^{60}Co activity mixed in 250 ml of water (each forming an aliquot). Each aliquot was deposited back onto the floodplain during a simulated flood. A total of 15 aliquots were deposited at representative locations to form a series of 'active plots' (hereafter referred to as AP1–15)), within the flood-zone, so that labelled sediment on each plot would be subjected to the potential remobilising effect of the overbank floodwaters, when the floodplain was next inundated. A further three plots were installed above the flood zone and used as controls (i.e. hereafter referred to as control plots (CP1–3)), to account for the loss or removal of labelled sediment, and hence, radioactive inventory, by processes other than by floodwaters. A simulated flood was created using a 300-mm diameter bottomless container which was embedded into the floodplain surface, so that a watertight seal was obtained. In order to recreate overbank flood conditions, one aliquot of labelled sediment was added to a further 25 l of water, vigorously shaken and then poured into the container. The water was allowed to infiltrate into the ground, thereby depositing the ^{60}Co-labelled sediment onto the floodplain surface (**Figure 4**). This approach was repeated for all 18 plots, with each simulated flood depositing a layer of ^{60}Co-labelled sediment to a depth of approximately 1.4 mm, which was representative of conditions in local rivers during typical flood events. The radiometric inventory associated with each plot was recorded before and after consecutive flood events using an in-situ gamma spectrometer. Data from the active plots over the three flood events were then correlated against the magnitude of each event, as indicated using peak river discharge, as well as two key spatial controls, namely the depth to which each active plot was inundated and their respective distances from the main river channel.

3.1.3. Key results and discussion

Before and after inventory values were corrected for radioactive decay and the proportion of the radiometric inventory lost due to processes other than by remobilisation by floodwaters, as indicated by the average of the three control plots, was removed from all 15 active plots. Minimum and maximum error bars were calculated at the 90% confidence level in order to

Figure 4. A bottomless container was embedded into the soil surface of a floodplain and 200 g of ^{60}Co-labelled sediment, mixed in 25 l of water, were poured into the container and allowed to infiltrate, thereby depositing the labelled sediment onto the soil surface under simulated flood conditions.

show the uncertainty associated with before and after measurements for each plot. At this particular juncture in the research, it was considered more meaningful, should similar work be conducted in the future on river floodplains elsewhere, if variations in initial inventories between plots were normalised and converted to an equivalent mass of sediment (expressed in g m^{-2}). Consequently, all plots started with an initial depositional mass of sediment equivalent to 1386 g m^{-2}. Where error bars between before and after measurements for each active plot do not overlap, the change in inventory is considered to be significant. This is interpreted as clear evidence that the labelled sediment was remobilised by overbank floodwaters. **Figure 5A–C** shows which plots recorded significant remobilisation over three consecutive flood events (i.e. termed the extent of remobilisation). They also indicate how much labelled sediment remained at each active plot after each flood event (i.e. termed the magnitude of remobilisation). For brevity, both the extent and mean magnitude, based on the mean of all 15 active plots, are summarised in **Tables 2** and **3**, respectively. Looking at data shown in **Table 2**, a clear inverse trend can be seen, which shows numbers of active plots that recorded significant remobilisation diminished with increasing time after initial deposition. Indeed, significant remobilisation ceased completely during the third inundation event at 119 days after initial deposition. There was neither correlation between the magnitude of each flood and the magnitude of remobilisation nor any correlation between magnitude of remobilisation and depth. A moderate yet significant direct correlation was recorded between magnitude of remobilisation and distance from river channel during the first flood event only. The direct relationship indicates that remobilisation intensified with increasing distance from the river. This finding, although

http://dx.doi.org/10.5772/intechopen.71304

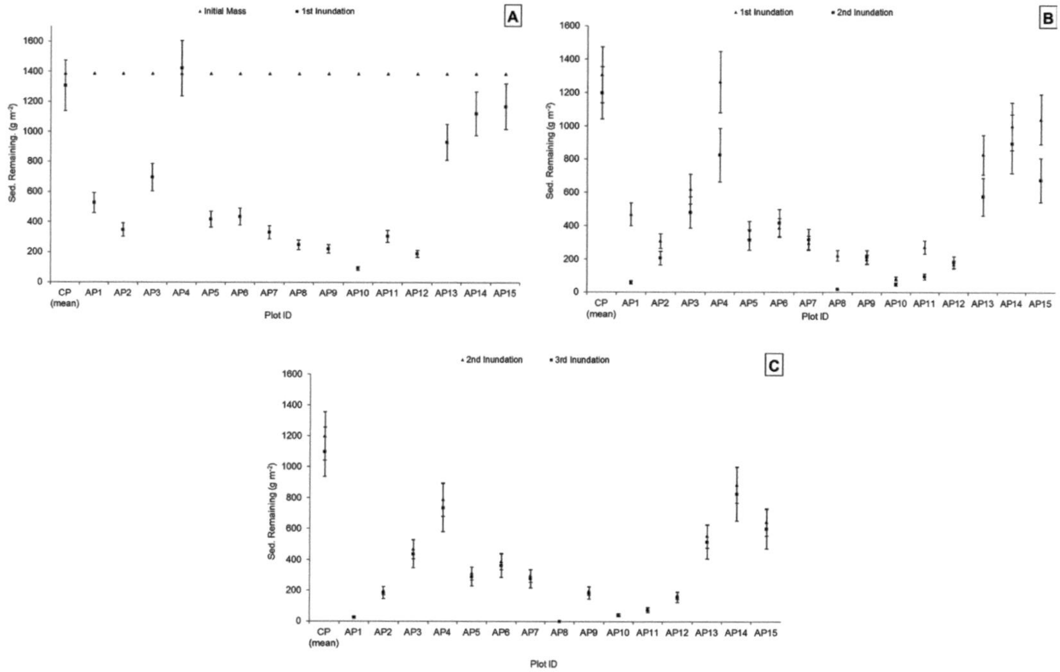

Figure 5. (A) Results of remobilisation after the first flood event. Error bars indicate the uncertainty associated with each measurement. Where error bars associated with before and after inundation values do not overlap, significant remobilisation of the ^{60}Co-labelled material has occurred at that location. Copyright © 2013 John Wiley & Sons, Ltd. (B) Remobilisation after the second flood event, although significant, was notably less than for the first flood event. Copyright © 2013 John Wiley & Sons, Ltd. (C) No significant remobilisation was recorded during the third flood event. Copyright © 2013 John Wiley & Sons, Ltd.

somewhat paradoxical, is attributed to the geometry and certain topographic features within this particular study site, both of which serve to temporarily divert but channel the fast-flowing floodwaters away from the river at the mid-point of the floodplain.

Interestingly, if data from **Table 2** are cross-referenced with data seen in **Table 3**, a number of key processes are revealed from which certain conclusions may be drawn. Firstly, flood-derived sediment appears to be extremely susceptible to remobilisation by subsequent floods during the first few weeks after deposition. Secondly, a significant amount of the same sediment

	Nos. of active plots	Inundation event		
		1st	2nd	3rd
% of active plots recording significant remobilisation	15	93 (14)	53 (8)	0
Time since initial deposition (days)	n/a	16	76	119

Nos active plots in parenthesis.

Table 2. The extent of remobilisation over three consecutive flood events, as indicated by ^{60}Co, and time since initial deposition for the River Taw floodplain.

Initial sed. mass (g m^{-2})	Mean remobilisation mass (g m^{-2})			Mean sed. mass remaining (g m^{-2})
	Event 1	Event 2	Event 3	
1386	884.63 (63.8)	165.06 (11.9)	24.27 (1.8)	312.04 (22.5)

Equivalent % values in parenthesis.

Table 3. The mean magnitude of remobilisation over three consecutive flood events, as indicated using ^{60}Co, and the mean amount of labelled sediment remaining after the third flood event for the River Taw floodplain.

also appears to remain vulnerable to remobilisation for up to 2 months after initial deposition. Thirdly, despite the average magnitude of remobilisation being insignificant at 1.8% during the third event (**Table 3**), an average of 22.5% of the initial inventory (equivalent to 312.04 g m^{-2}) remained after the third event. This is interpreted as evidence that the remaining portion of the flood-deposited (^{60}Co-labelled) sediment gradually stabilised. This stabilising effect is attributed to the material gradually becoming incorporated into the main soil profile and entering into temporary storage. This not only effectively made it unavailable for further remobilisation but also storage occurred almost 4 months after the sediment was initially deposited.

3.1.4. *Uncertainties*

A number of key uncertainties need to be acknowledged. Arguably, the most profound of which is that ^{60}Co is largely untried and untested as a fine-sediment tracer, and thus, the resultant data remain uncorroborated. Three independent approaches were undertaken in an attempt to rectify this uncertainty. The first approach involves presenting data from a parallel study using a different artificial radionuclide, namely Cs-134 (^{134}Cs; $t_{0.5}$ = 2.06 years). Sediment labelled with ^{134}Cs was duly deployed in exactly the same manner described above but on another floodplain in a different river system. Although the flow regime of the two rivers and the floodplain characteristics differ markedly, equivalent remobilisation values and general timelines from data recorded using ^{134}Cs over three flood events in **Tables 4** and **5** are striking in their similarity with data from **Tables 2** and **3** using ^{60}Co. The second approach involved performing independent sorption tests on surface soil taken from each floodplain in order to determine whether the response of both radionuclides to provide information on remobilisation dynamics was comparable. Textural characteristics on surface soils taken from each floodplain (**Table 6**) indicated that <63 μm fraction (i.e. silt and clay) was generally similar and represented ca. 80–85% of each sample, as were other key indicators, including clay fractions

	Nos. of active plots	Inundation event		
		1st	2nd	3rd
% of active plots recording significant remobilisation	15	87 (13)	47 (7)	0
Time since initial deposition (days)	n/a	50	74	104

Nos active plots in parenthesis.

Table 4. The extent of remobilisation over three consecutive flood events, as indicated using ^{134}Cs, and time since initial deposition for the River Culm floodplain.

Initial Sed. Mass (g m^{-2})	Mean remobilisation mass (g m^{-2})			Mean sed. mass remaining (g m^{-2})
	Event 1	Event 2	Event 3	
1386	687.17 (49.6)	173.03 (12.5)	37.94 (2.7)	487.82 (35.2)
Equivalent % values in parenthesis.				

Table 5. The mean magnitude of remobilisation over three consecutive flood events, as indicated using ^{134}Cs, and the mean amount of labelled sediment remaining after the third flood event for the River Culm floodplain.

(<2 μm), median particle size (D_{50}) and SSA (**Table 6**). Sorption tests were undertaken on 12 soil samples from both floodplains according to Sparks [72], six samples of which were labelled with ^{60}Co and six labelled with ^{134}Cs, at the same activity concentration. Samples were then allowed to rest for 24 h, and then the solid and liquid fractions of each aliquot were assayed together. After decanting off the supernatant (liquid) fraction, both fractions were then re-assayed separately. After averaging individual values for each group, results indicated for both radionuclides that the solid (i.e. sediment) fraction retained 99.8% of the applied inventory. The third method of corroboration involved undertaking a 'paired-plot' approach, with the aim of comparing the response of both radionuclides under similar floodwater-flow conditions. Two active plots were established 0.15 m apart (measured from perimeter to perimeter) on a uniform area of one of the floodplains, one installed using ^{134}Cs-labelled sediment and the second installed using ^{60}Co-labelled sediment. An assumption was then made that floodwater-flow conditions across flat topography with sparse vegetation cover would be sufficiently similar over short distances [80, 81] to allow results from each set of results to be compared. Four sets of paired plots were established (hereafter referred to as PP1-PP4), all within a 6 m radius of each other, and all plots were duly inundated during the same flood event, to a depth of ca. 0.15 m. The results, which are shown in **Table 7**, reveal that ^{60}Co consistently records a slightly lower magnitude of remobilisation in comparison to data generated using ^{134}Cs. Differences range from a negligible 0.5% (PP4; ^{60}Co < ^{134}Cs) to a notably larger 12.5% (PP2; ^{60}Co < ^{134}Cs), with a mean difference of 6.5% (^{60}Co < ^{134}Cs; n = 4). One possible reason for the contrasting values recorded by PP2 may be due to spatially variable flow conditions at that location. However, if this value is treated as an outlier and excluded, the average difference between the two radionuclides falls to just 4.5% (^{60}Co < ^{134}Cs; n = 3).

	River Taw floodplain soil	River Culm floodplain soil
>63 μm (%)	14.6	20.8
<63 μm (%)	85.4	79.2
<2 μm (%)	6.8	4.7
Median D_{50} (μm)	22	28
Specific surface area (SSA) (m^{-2} g^{-1})	1.31	1.40
Bulk density (g cm^{-1})	1.66	1.70

Table 6. Key characteristics associated with surface soil from the River Taw and River Culm floodplains.

	^{134}Cs		^{60}Co		% Difference
	Mass remaining ($g\ m^{-2}$)	% Remaining	Mass remaining ($g\ m^{-2}$)	% Remaining	
PP1	1191.93	86	1264.51	91.2	5.2
PP2	902.36	65.1	1075.92	77.6	12.5
PP3	1032.28	74.5	1140.39	82.3	7.8
PP4	1001.44	72.3	1009.64	72.8	0.5
Average	1032.00	74.5	1122.62	81.0	6.5

Table 7. Results from a paired-plot approach to determine differences in ^{60}Co and ^{134}Cs response under similar remobilisation conditions during one flood event.

3.2. Example 2: dispersed earthworm casts as a sediment-source

3.2.1. Experimental aim

The aim of this investigation was, again, to exploit the high sorption characteristics of ^{60}Co by testing its tracing capabilities on a biologically driven sediment detachment mechanism that would be particularly challenging for most other forms of tracer. The underlying rationale for this work was to confirm or refute the implication that fine sediment from disintegrating earthworm casts contributes to agricultural run-off. Previous attempts to quantity this process have met with only limited success [82–84]; however, due to the challenging nature of the erosion mechanism under investigation and of inherent difficulties of tracing objects that start as tangible 3-dimensional structures, but when weathered and disintegrated, lose their physicality by reverting back to fine sediment. Consequently, a prototype technique for labelling intact earthworm casts was designed in order to attempt to document the mobility of dispersed earthworm casts on vegetated hill slopes. More detailed background information on this particular tracing approach can be accessed in Greenwood [16]. Motivation for attempting a tracing investigation of this nature revolves around the fact that earthworms are prevalent in all edaphic environments, except deserts and permafrost regions. Anecic, or deep-burrowing, species are responsible for the vertical movement of soil due to a number of fundamental lifecycle traits namely burrowing and feeding. Although both activities are generally acknowledged to beneficially modify the soil and improve soil structure through mixing and incorporating surface organic matter into the upper soil horizon, a by-product of those activities is the production of earthworm casts on the soil surface (**Figure 6**). Researchers have thus speculated that the presence of surface casts not only represents an on-going soil detachment mechanism but also inadvertently renders casts vulnerable to weathering by impacting raindrops and leaves the resultant fine sediment susceptible to mobilisation by surface run-off. It is possible, therefore, that earthworm casts could represent a substantial sediment source. Given the ubiquity of anecic earthworms on vegetated hill slopes in temperate environments, along with their constant burrowing and feeding, it is likely that notable quantities of fine sediment and sediment-bound nutrients may remain in a constant state of flux, particularly on areas, where earthworms are particularly active and surface casts are abundant.

Figure 6. The on-going production of casts makes the sediment vulnerable to weathering and possible downslope mobilisation, even on relatively gentle slopes.

3.2.2. Method

The initial method for labelling intact earthworm casts with ^{60}Co was based on a technique originally reported by Toth and Alderfer [55] for labelling WSAs by immersion into a known activity of ^{60}Co mixed with water. Small groups of casts were air dried, weighed, and then immersed 'en masse' into a shallow tray containing a known solution of ^{60}Co mixed with a further 2 l of water. After immersion for a predetermined period of time (usually around 90 seconds), casts were removed, allowed to drip for ca. 30 s and then re-weighed (wet). The increase in mass (g) was multiplied by the original activity concentration (Bq ml^{-1}) of the radionuclide solution to derive an initial inventory for each group of casts. Labelled casts were then deployed on a 0.15 m^2 area of prepared grassland located at the hill slope divide of a ca. 10% slope (5.7°). Vegetation within the plot was trimmed, and each side of the plot was bounded using an aluminium sheet embedded into the soil surface to prevent run-on. A metal channel was embedded at the downslope end of the plot, flush with the soil surface, to guide any run-off and run-off-eroded sediment into an awaiting container located in a nearby reception pit (**Figure 7**). Any unlabelled casts within the prepared plot were removed, and 15 ^{60}Co-labelled casts, representing 190.7 g of sediment and with a group inventory of 224.3 Bq (equivalent to 1.18 Bq g^{-1}), were evenly spread across the upper part of the erosion plot. Casts were exposed to natural weather events over the following 76 days. Monitoring generally took place on a weekly basis, but the frequency of site visits was ultimately driven by the occurrence of rainfall events. During the operational phase, 186.3 mm of rainfall was recorded which generated 16 run-off samples. Sediment was separated from the run-off and

Figure 7. A number of ^{60}Co-labelled earthworm casts were spread across a bounded area of pasture and exposed to natural weather events for 76 days.

radiometrically assayed on a laboratory-grade gamma spectrometer, and a mass balance approach was used to partition the relative proportions of run-off-derived ^{60}Co-labelled sediment from any unlabelled sediment.

3.2.3. Key results and discussion

Data presented in **Table 8** show that of the 190.7 g of sediment originally deployed across the erosion plot in the form of 15 ^{60}Co-labelled earthworm casts, 26.82 g of the labelled sediment (equivalent to 14.1% of the original mass) was removed from the plot by surface run-off. In descriptive terms, the largest sample mass was 9.29 g and was recorded on day 5, the lowest sample mass was 0.14 g and was recorded on day 61 and the average mass based on all 16 samples was 1.68 g. The average mass for the first eight samples recovered from day 5–30 was 2.51 g, whereas the average mass for the latter eight samples recovered between days 32–76 was 0.85 g. **Figure 8** illustrates temporal variations in sediment recovery. Higher sediment recovery rates during the early stage of the investigation probably reflect higher sediment

	^{60}Co (travel distance 0.1–0.3 m)	Unlabelled
Total initial sed. mass (g)	190.7	n/a
Total run-off removed (g)	26.82	59.51
% Removed	14.1	n/a
% Retained	85.9	n/a

Table 8. The proportion of ^{60}Co-labelled earthworm-derived sediment removed by surface run-off and the proportion retained.

availability. In contrast, lower recovery rates during the latter half of the experiment could either indicate the gradual exhaustion of the available (labelled) sediment supply, or it could indicate stabilisation and incorporation of the remaining labelled material back into the soil profile. Despite the gradual reduction in labelled sediment recovery, data seen in **Figure 8** clearly highlight that sediment from dispersed earthworm casts can act as a notable sediment source for at least a few months after initial production. Interestingly, unlabelled sediment, presumably generated after deployment of the labelled casts, was sporadically recovered on days 23 and 27 and then generally continuously recovered from day 38 onwards. If the unlabelled material comes from fresh earthworm casts, then the initial absence may reflect the fact that all unlabelled casts were removed from the plot when the investigation commenced. Similarly, the sporadic appearance of unlabelled sediment on days 23 and 27, followed by a generally sustained presence from day 38 onwards, may reflect a steady increase in unlabelled cast production by the resident earthworm population within the vicinity of the erosion plot.

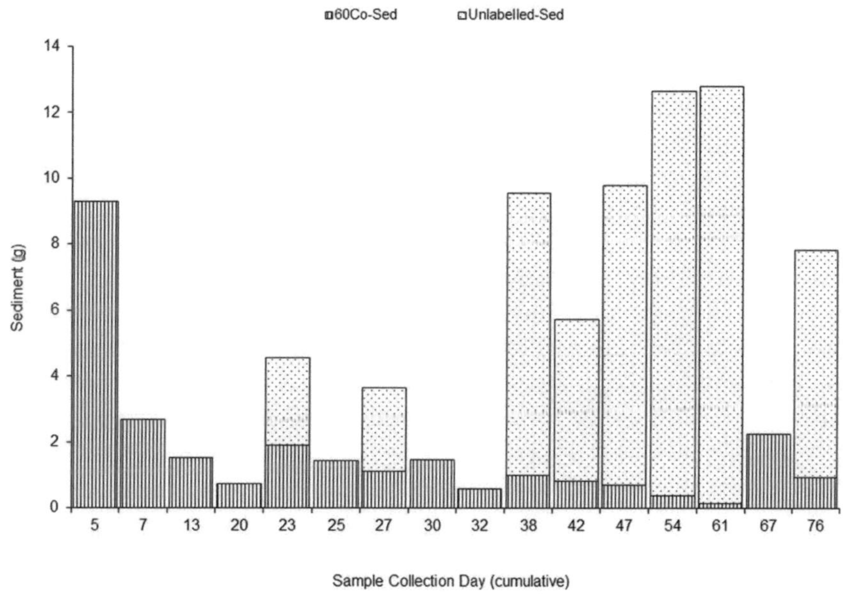

Figure 8. Results of an in-situ erosion experiment showing temporal variations in the amount of both ^{60}Co-labelled and unlabelled sediment recovered in surface run-off over the 76 day operational period.

3.2.4. *Uncertainties*

While the results presented above provide 'proof of concept' and support the idea that the tracing approach is generally sound and certainly worth developing, there are a number of issues and uncertainties that need to be addressed. One of the most pressing is the way in which initial inventories were calculated. Since a failure to determine an accurate initial inventory will generate erroneous data without an appropriate correction factor, initial refinements should focus on ways in which an initial group inventory can be established with more certainty. The original assumption predicted that dried, intact earthworm casts will sorb an amount of radionuclide that is proportional to the original concentration. Subsequent sorption tests were undertaken to test this hypothesis on small groups of earthworm casts using ^{60}Co, as well as ^{134}Cs, again, in order to assess and compare their respective sorption behaviours. A total of six aliquots were prepared, three containing ^{60}Co concentrations equivalent to 0.25, 0.5, and 1.0 Bq ml^{-1}, each in 400 ml of water, and three containing ^{134}Cs at similar concentrations and volumes. Earthworm casts were oven dried and then sorted into 18 groups, with each group consisting of five similar-sized casts, with an average weight of 39.8 g. The labelling procedure firstly involved measuring the radionuclide activity of a 100-ml sub-sample of radionuclide solution from each aliquot. After measuring, each sub-sample was then merged back into its respective aliquot, and the first group of casts was immersed into an aliquot for 4 min (hereafter referred to as Dip 1), removed and allowed to drain for ca. 30 s. A 100 ml sub-sample of the remaining liquid was, again, taken from each aliquot and analysed to determine the change in activity. This approach was repeated twice more (i.e. Dip 2 and Dip 3) until three groups of casts had been immersed into each aliquot. This refined approach follows the premise that the change (i.e. reduction) in inventory concentration after immersion of a group of casts must solely be due to the casts themselves. This technique is believed to represent a more accurate way of establishing a precise inventory for each group. Preliminary results from the refined sorption tests are shown in **Figures 9** and **10** for ^{60}Co and ^{134}Cs, respectively, and suggest that the original assumption may not necessarily be correct for contrasting reasons. By way of explanation, a group of casts recording zero along the Y-axis would signify sorption at an equivalent concentration as the original solution (i.e. no net loss or gain). Conversely, positive values along the Y-axis signify that a group of casts sorbed more than the available inventory concentration, whereas negative values signify that casts sorbed less than the available inventory concentration. For ^{60}Co, three groups of casts did indeed sorb more than the original concentration, whereas six groups sorbed notably less than the available inventory. For ^{134}Cs, one group sorbed slightly less than the available inventory concentration, whereas eight groups sorbed notably more than the available concentration. While these findings suggest sediment recovery data presented in **Figure 10** may be underestimated, by-products of the recent sorption tests are 18 groups of casts, each with a precise inventory. Consequently, scope exists to utilise those groups of casts experimentally by undertaking a series of replicate erosion experiments to further explore the potential of this particular tracing technique.

3.3. Example 3: livestock-poached pasture as a sediment source

3.3.1. *Experimental aim*

In the previous two examples presented, labelling was undertaken by exposing small quantities of fine-sediment to quantities of open-source ^{60}Co; whereupon the resultant labelled material

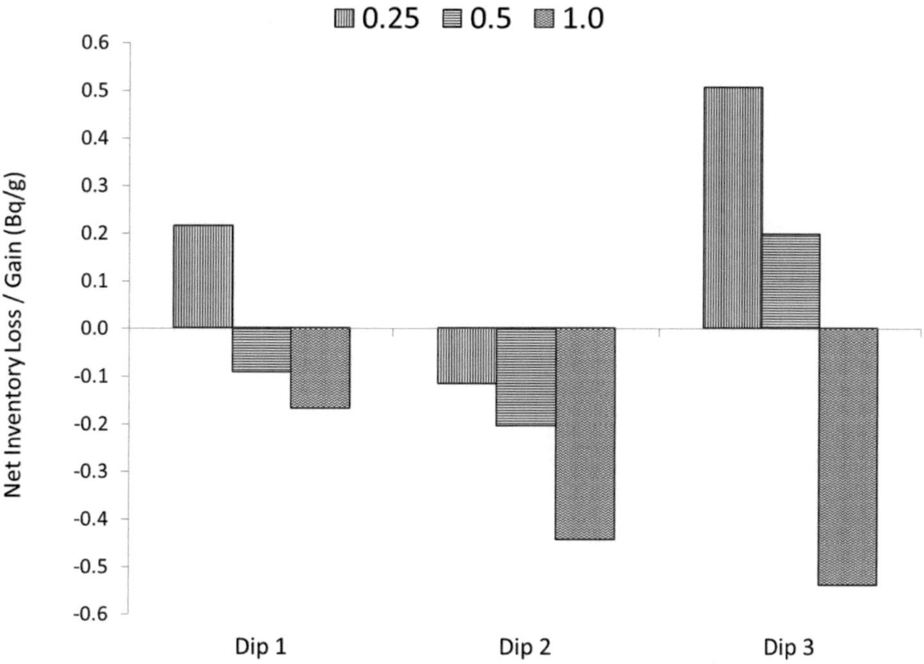

Figure 9. Results of sorption tests on small groups of intact earthworm casts by immersion into known activities of ^{60}Co, each mixed in 400 ml of water.

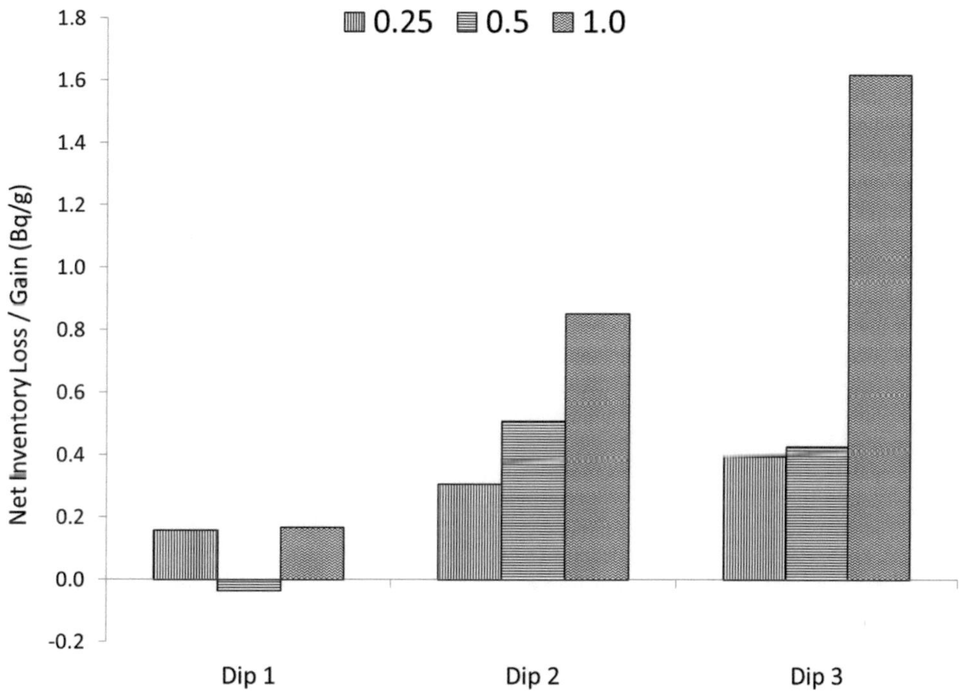

Figure 10. Results of sorption tests on small groups of intact earthworm casts by immersion into known activities of ^{134}Cs, each mixed in 400 ml of water.

was redeployed, albeit in different settings and under different erosion scenarios. In this final example, labelling was undertaken in-situ across six 0.2 × 0.2 m areas of grassland that had been heavily trampled and de-vegetated by cattle over the winter period, which left the soil with a slurry-like consistency, in a process known as livestock poaching (**Figure 11**). Again, more detailed background information on this particular tracing approach can be accessed in Greenwood [16] and in Greenwood et al. [58]. In brief, poaching is a common trait associated with domestic herd animals that frequently occur within confined areas of intensively managed grasslands, when the herd repeatedly congregates in certain areas. This can include field gateways, around feed-mangers and water-troughs, or along the banks of unfenced watercourses. If left unchecked, the repeated churning action of their hooves tramples and mixes the soil with any surface water. Vegetation cover is quickly depleted, and the soil can take on a slurry-like consistency with poor structure and limited cohesion. In this state, it is particularly vulnerable to redistribution by, for instance, impacting raindrops, and/or by surface run-off. Against this background, advances and recent developments in catchment-scale sediment fingerprinting techniques [17, 85] have demonstrated that intensively managed grasslands may contribute more fine sediment and sediment-associated nutrients and contaminants than originally presumed. Given legislation such as the WFD that aims to force EU member states to improve water quality in rivers and streams, a considerable number of studies have sought to identify potential sediment sources within intensively managed grasslands [10, 86, 87]. Resulting from those studies, poaching by livestock is been cited as a likely source of eroded sediment [38]. However, little work has been undertaken to quantify the process, again, due

Figure 11. Areas in fields where livestock repeatedly congregate, such as gate entrances, often become heavily poached; causing the surface soil to lose its cohesion and ability to resist erosion. In intensively managed grassland systems, areas such as this may act as sediment-sources.

to the dearth of available techniques capable of accurately documenting the movement of very small quantities of fine sediment at the required spatial and temporal scales [88]. Motivated by an urgent need to quantify this process and challenged by the complexities of the erosion mechanism and the very short timescales over which it can potentially operate, an in-situ method of labelling poached areas with ^{60}Co, for the purpose of documenting the extent to which sediment might be subsequently redistributed, was devised and executed.

3.3.2. Method

Small areas (0.04 m^2) of livestock-poached soil were each directly labelled with 1040 Bq of ^{60}Co activity mixed in 250 ml of water, using a correspondingly sized drip-type rainfall simulator, which was able to distribute an even supply of the radionuclide material over each area under controlled conditions. Nine areas were labelled within a 2.6 ha. field using this approach; three areas were established within the poached gateway seen in **Figure 11** on slopes ranging from 4.5 to 8% (2.6–4.6°), three were established along the bank of a poached watercourse on slopes ranging from 6 to 21.3% (3.4–12°), and three were established on a nearby flat, fully vegetated and un-poached area for use as controls. After labelling, each area was precisely marked and then allowed to rest for 24 h to allow the material to fully infiltrate and sorb to the soil before an initial surface inventory was established. Again, exploiting the high sorption capacity of ^{60}Co to fine sediment, a pilot study was undertaken 1 week before installation of the nine areas, which indicated that ca. 90% of the applied inventory remained within the upper 25–30 mm of an undisturbed soil profile. Initial surface inventories were established at 16 predetermined measurement points across each area using the in-situ gamma spectrometer previously seen in **Figure 3B**. An analytical protocol was devised to ensure that consistency was maintained throughout the measurement procedure, which coupled with undertaking measurements at the same 16 predetermined points, ensured the same ca. 86% of each area was measured every time. After installation, the nine areas were thus operational for a total of 65 days over the winter period, 2007. During this time, all nine areas were re-measured on three separate occasions after one or a series of natural rainfall events. Changes in surface inventories were determined relative to previous measurements, and these were used as a basis for inferring sediment redistribution across each poached area. On terminating the experiment, areas used as controls were shallow cored, sectioned in 3 mm intervals and assayed on a laboratory-grade gamma spectrometer to establish the vertical distribution of ^{60}Co within the soil profile. This information was then incorporated into a calibration procedure and used to convert changes in radiometric values to equivalent soil redistribution values.

3.3.3. Key results and discussion

Due to this investigation still being at a preliminary stage, data are presented in **Table 9** from one poached area only over the three re-measurement campaigns in order to provide proof-of-concept and to tentatively demonstrate the viability of the tracing technique. Radiometric values are decay and control plot corrected, and resultant erosion values are presented in kg m^{-2} d^{-1}, with ± uncertainty values calculated at the 95% confidence level. In instances where the surface inventory was less than the previous value, this is interpreted as evidence of

Re-measurement 1

	Column 1		Column 2		Column 3		Column 4		
	Redist.	**± Error**	**Redist.**	**± Error**	**Redist.**	**± Error**	**Redist.**	**± Error**	**Mean (kg m−2 d−1)**
	(kg m^{-2} d^{-1})		**(kg m^{-2} d^{-1})**		**(kg m^{-2} d^{-1})**		**(kg m^{-2} d^{-1})**		
Row 1	−0.88	0.15	−2.20	0.29	dep	n/a	−1.27	0.24	
Row 2	−0.14	0.03	−1.31	0.22	−1.10	0.14	dep	n/a	
Row 3	−0.36	0.06	−0.46	0.08	dep	n/a	−1.03	0.17	
Row 4	−2.14	0.41	−2.51	0.44	−1.68	0.30	dep	n/a	
Mean (kg m^{-2} d^{-1})									−1.26

Re-measurement 2

	Column 1		Column 2		Column 3		Column 4		
	Redist.	**± Error**	**Redist.**	**± Error**	**Redist.**	**± Error**	**Redist.**	**± Error**	**Mean (kg m−2 d−1)**
	(kg m−2 d−1)		**(kg m−2 d−1)**		**(kg m−2 d−1)**		**(kg m−2 d−1)**		
Row 1	−1.15	0.27	−1.08	0.24	−1.15	0.20	−0.93	0.19	
Row 2	−1.33	0.24	−0.46	0.07	−1.38	0.21	dep	n/a	
Row 3	−0.95	0.28	−0.54	0.14	−1.17	0.33	−1.54	0.46	
Row 4	−1.23	0.26	−0.79	0.15	−2.25	0.61	dep	n/a	
Mean (kg m^{-2} d^{-1})									−1.14

Re-measurement 3

	Column 1		Column 2		Column 3		Column 4		
	Redist.	**± Error**	**Redist.**	**± Error**	**Redist.**	**± Error**	**Redist.**	**± Error**	**Mean (kg m−2 d−1)**
	(kg m−2 d−1)		**(kg m−2 d−1)**		**(kg m−2 d−1)**		**(kg m−2 d−1)**		
Row 1	−1.27	0.34	−0.83	0.24	−0.69	0.18	−0.58	0.12	
Row 2	−1.65	0.29	−0.61	0.11	−1.01	0.24	dep	n/a	
Row 3	−1.62	0.42	−1.49	0.36	−0.99	0.28	−0.99	0.27	
Row 4	−1.14	0.22	−1.00	0.18	−1.75	0.41	dep	n/a	
Mean (kg m^{-2} d^{-1})									−1.11

Table 9. Soil redistribution values measured over three separate re-measurement campaigns spanning 65 days.

the removal of labelled sediment at that particular location. In contrast, where the surface inventory was greater than the previous value, this is interpreted as evidence of the deposition of labelled sediment at that location. Due to the complexities of calculating deposition, however, no data are currently presented for those points. Despite their absence, however, evidence of deposition was only recorded at four points during the first re-measurement and at two points each during the second and third re-measurements. Consequently, the predominant redistribution mechanism over the three re-measurement campaigns, certainly at this particular poached area, was clearly erosion; signifying a net loss of material from within the confines of the monitored area during the 65 days the experiment was operational. Average erosion values over the three re-measurement campaigns were 1.26, 1.14 and 1.11 kg m^{-2} d^{-1}, respectively. The stepwise reduction in net soil loss over the three re-measurement campaigns is thought to reflect the poached soil gradually drying, regaining its structure and offering increasing resistance against redistribution processes as weather conditions improved. In addition to this, the gradual regrowth and recovery of grass probably exerted an increasingly stabilising influence on the soil over time. These very tentative yet encouraging findings, albeit from one location only, support the original hypothesis that livestock-poached areas can act as a sediment source in intensively-managed grassland systems. It is worth emphasising, however, that as poaching more often occurs as isolated islands surrounded by a sea of dense grassland, it is highly likely that any eroded material would be trapped by the surrounding grass and probably quickly reincorporated back into the soil matrix. However, on poached areas lying immediately adjacent, or close to watercourses, it is possible that the eroded material could migrate into the watercourse and contribute to reducing water quality downstream of its entry point.

3.3.4. *Uncertainties*

Soil redistribution depths were determined using one of many available calibration procedures, each of which converts inventory values to equivalent soil redistribution values in a slightly different way. Depending on which calibration procedure is chosen, therefore, redistribution rates may vary, even when based on the same radiometric data. As ^{60}Co is still largely untried and untested as a fine-sediment tracer, and the tracing approach is not only novel but also is still in its infancy, a logical way of off-setting at least some of the uncertainty associated with the resultant redistribution data would be to process the same radiometric data using two different calibration models. Such an approach would permit data sets to be compared and provide an indication of the degree of uncertainty involved [89].

4. Problems with using Cobalt-60

Despite the numerous benefits and advantages that have been highlighted in previous sections regarding open-source ^{60}Co as a fine-sediment tracer, during the course of undertaking these, and other, sediment tracing investigations using radioactive ^{60}Co, a number of problems were encountered. One of the most prominent was initially locating a supplier of open-source radioactive material and thereafter obtaining/ensuring that the necessary licences and permits were in-place to enable the radioactive material to be purchased, stored, used and, where applicable, disposed of. Also important was the availability of trained personnel qualified to partition and dispense stock material at activities suitable to undertake experimental work. Also paramount when attempting to develop novel tracing techniques is to have ready access to highly specialised, and expensive, in-situ and laboratory-based gamma spectrometers so that analyses can be performed under a range of environments, sometimes more than once. Also problematic was finding sympathetic landowners or farm managers that were willing to allow work to be conducted on their land, largely due to the stigma associated with radioactivity. This particular issue admirably illustrates the contentiousness associated with radiation *per se*, despite it being very low level and especially with the deliberate, and arguably contentious, release of such material into the environment, particularly for the purpose of studying what many members of the general public would probably view as a trivial subject with limited real-world benefits.

5. Key refinements and prospects for future work

In terms of key refinements and prospects for future work, with regard to Example 1, rather than using intuition and experience to select sites on which to install active plots, more representative spatial coverage could be obtained by adopting a more focused approach where, for instance, plots are installed using an equidistant grid system, with spacings that are appropriate to the size of the floodplain under scrutiny. In addition, a progressive reduction in the analytical precision of most active plots was noted over the second, but particularly over the third, inundation event. This was attributed to diminishing radiometric signals emitted by the finite supply of labelled sediment as supply gradually became exhausted, and/or high uncertainties due to the necessarily short counting time (600 s) devoted to each analysis. Both effects could be mitigated in several ways. One way is to increase the amount of labelled sediment and associated inventory deposited at each plot. Alternatively, a simpler, somewhat less contentious and more obvious way is to extend the counting time per assay. While this would initially appear to be a good alternative, since flooding occurs almost exclusively in the autumn/winter period, a time when natural daylight hours are around 6–7 h only, this would necessarily extend the total analytical time from 3 days to probably in excess of 4 days per floodplain site (i.e. >8 days in total). Undertaking analyses over timescales such as this and taking into account the time needed for floodwaters to recede, there exists a good chance that multiple floods could occur in quick succession before further analyses could be performed, thereby resulting in a loss of information on the extent and magnitude of remobilisation over single flood events.

With regard to Example 2, in light of the refined labelling approach and presumed increased precision, just a flavour of the possible tests that could be undertaken in future might include using simulated rainfall and erosion plots lined with turf to simulate permanent pasture. Such an approach could offer a convenient and precise way of looking at key controlling factors, such as slope and vegetation density, and the degree to which they influence rates of sediment recovery and the travel distance of the eroded material. Further work could also be undertaken on simulated cultivated soil; again, using artificial rainfall and erosion plots; but in this instance, to not only determine travel distance of the dispersed sediment from earthworm casts on unconsolidated soils, but perhaps also study the rate of removal from crusted soils. While the use of both artificial rain and erosion plots is a convenient way of undertaking experiments, neither approach can substitute for the effect that time has on weathering and erosion, however. Consequently, it could also be worth deploying groups of casts in situ on natural pasture and use natural rainfall as the principle weathering agent, in a similar manner to that described earlier and monitor over several months or until the radiometric signal associated with run-off-recovered sediment falls to below detection levels. As a further addendum to this basic idea, it might also be worthwhile installing in-situ plots, bounded and spaced relatively close to each other, but each with a progressively longer distance from the area of deployment to the channel outlet in order to explore the relationship that space and time has on the travel distance of eroded casts.

With regard to Example 3, in light of the tentative yet encouraging results presented earlier, a number of changes could be made that could reduce the uncertainty. Firstly, rather than installing areas based on observations and intuition, a denser network of systematically placed plots across the full extent of each poached area would generate more representative spatial information. As for the same reasons above, analytical counting times were deliberately kept short due to lack of daylight hours when the experiment was done. While extending them would reduce the uncertainty associated with individual redistribution estimates, it would also mean that total analytical times would be commensurately longer. Consequently, a trade-off would need to be made between longer count times, resulting in increased accuracy and completing the analyses in a realistic period of time. One additional refinement to the existing technique that would be adopted in future would be to label poached areas and ascertain initial surface inventories, but re-measure once only and then abandon, in favour of installing new sites on different poached areas elsewhere. Reasons for this seemingly drastic approach relate to the sensitivity of the mathematical conversion procedure, and the fact that its ability to convert changes in surface activities to equivalent soil redistribution depends on the distribution of the radionuclide declining exponentially with depth. While this relationship remained valid immediately after labelling, the progressive redistribution of labelled surface material, along with the likely ingression and mixing of unlabelled sediment from elsewhere, would probably cause the exponential relationship to steadily break down. Although the timescale over which this would occur are unknown, the end result would be increasingly erroneous data.

6. Concluding remarks

The chapter has described and reviewed three prototype sediment tracing approaches using the artificial gamma-emitting radionuclide ^{60}Co, each of which was designed to measure the redistribution, remobilisation or dispersal of fine-sediment under a different scenario. The first example looked at the fate of flood-deposited sediment on river floodplains and found that the material was very susceptible to remobilisation during the first flood event immediately after deposition. However, it also found that a significant proportion was still mobile for up to 2 months after deposition, after which, whatever remained was incorporated into the bulk soil profile and stored. The second example presented in the chapter sought to determine whether dispersed earthworm casts act as a sediment source on vegetated hill slopes. The study found that sediment from dispersed casts did indeed act as a sediment source, and importantly, it also remained mobile for at least a few months after initial production. The final example sought to determine the extent to which areas of livestock-poached pasture were subjected to redistribution processes. The study found that such areas did indeed experience sediment redistribution and that erosion, or a net loss of fine-sediment from poached areas, was the predominant process. The above findings, although preliminary, are not only sufficiently encouraging to warrant further refining each approach using ^{60}Co but also collectively represent essentially unique information that contributes to an improved understanding of some of the widespread yet largely under-studied sediment mobilisation and transfer mechanisms that commonly operate within catchments. While some limitations and issues were duly noted with ^{60}Co, most of those could be readily circumvented with sufficient advanced planning. Indeed, it is therefore difficult to imagine how information gained from the three experimental examples presented above could be obtained using the more common tracers reviewed earlier in the chapter. Based on these findings, the versatility of ^{60}Co as a fine-sediment tracer has been confirmed, and its efficacy to provide accurate erosion data in challenging settings has undoubtedly expanded the overall scope and diversity of sedimentation scenarios that can currently be investigated. With additional work to explore and further develop the capabilities of ^{60}Co, it might eventually be added to the existing suite of tracers and tracing techniques currently available to researchers studying sedimentation processes.

Acknowledgements

Some of the research referred to in this chapter was generated by two grants that were awarded to the author. The first is a Natural Environment Research Council (NERC)-funded PhD. studentship (Ref: NER/S/A/2004/12291A), and the second is a Short Term Scientific Mission (STSM) (COST-STSM-ES1306-34793), which forms part of the European Union-funded COST ES 1306 Initiative, Connectivity in Geomorphology. Thanks are extended to Mr. J. Snell of Newnham Barton Farm, north Devon, UK, on whose land some of the research presented here was undertaken. I would also like to acknowledge John Wiley Publishers Ltd. for granting copyright permission to reproduce certain figures and tables (Licence No. 4151340122828).

Author details

Philip Greenwood

Address all correspondence to: philip.greenwood@unibas.ch

University of Basel, Basel, Switzerland

References

[1] Barrow CJ. Environmental Management: Principles and Practise. London: Routledge; 2000

[2] Zapata F. The use of environmental radionuclides as tracers in soil erosion and sedimentation investigations: Recent advances and future developments. Soil & Tillage Research. 2003;**69**:3-13

[3] Walling DE. The changing sediment loads of the world's rivers. In: Sediment Dynamics in Changing Environments. International Association of Hydrological Sciences (IAHS), Publication No. 325. 2008. p. 323-338

[4] Walling DE, Quine TA. The use of caesium-137 measurements in soil erosion surveys. In: Erosion and Sediment Transport Monitoring Programmes in River Basins. International Association of Hydrological Sciences (IAHS), Publication No. 210. 1992. p. 143-152

[5] El-Swaify SA, Flach KW. Foreword. In: Lal R, editor. Soil Erosion Research Methods. Iowa: USA: Soil and Water Conservation Society; 1988

[6] Brown LR. Conserving soils. In: Brown LR, editor. State of the World. Vol. 53. New York: Norton; 1984, 1984. p. 53-75

[7] Imeson A. Desertification, Land Degradation and Sustainability: Paradigms, Processes, Principles and Policies. UK: Wiley-Blackwell Publishing; 2012

[8] Cherlet M, Ivits E. Compilation of a new atlas of desertification and contribution to a global assessment. In: de Boever M, Khlosi M, Delbecque N, de Pue J, Ryken N, Verdoodt A, Cornelis W, Gabriels D, editors. Desertification and Land Degradation: Processes and Mitigation. Ghent University, Belgium: UNESCO Chair of Eremology; 2013

[9] Gibbs HK, Salmon JM. Mapping the world's degraded lands. Applied Geography. 2015;**57**:12-21

[10] Evans R, Collins AL, Zhang Y, Foster IDL, Boardnam J, Sint H, Lee MRF. A comparison of conventional and ^{137}Cs-based estimates of soil erosion rates on arable and grassland across lowland England and Wales. Earth-Science Reviews. 2017;**173**:49-64

[11] Walling DE, Collins AL. The catchment sediment budget as a management tool. Environmental Science & Policy. 2008;**11**:136-143

[12] Pulley S, Foster I, Antunes P. The application of sediment fingerprinting to floodplain and lake sediment cores: Assumptions and uncertainties evaluated through case studies in the Nene Basin, UK. Journal of Soils and Sediments. 2015;**15**(10):2132-2154

[13] Europa 2017. The EU Water Framework Directive http://ec.europa.eu/environment/water/water-framework/info/timetableen.htm [Accessed 25/07/17]

[14] Pulley S, Van der Waal B, Collins AL, Foster IDL, Rowntree K. Are source groups always appropriate when sediment fingerprinting? The direct comparison of source and sediment samples as a methodological step. River Research and Applications. 2017:1-11

[15] Walling DE. Tracing versus monitoring: New challenges and opportunities in erosion and sediment delivery research, Chpt. 2. In: Owens PN, Collins AJ, editors. Soil Erosion and Sediment Redistribution in River Catchments: Measuring, Monitoring and Management. Oxfordshire, UK: CABI Publishing; 2006

[16] Greenwood P. Applications of active tracer techniques in soil erosion and catchment sediment source investigations [unpublished PhD thesis]. University of Exeter, UK: Department of Geography; 2010 https://www.researchgate.net/publication/303494922_Applications_of_Active_Tracer_Techniques_in_Soil_Erosion_and_Catchment_Sediment_Source_Investigations

[17] Collins AL, Pulley S, Foster IDL, Gellis A, Porto P, Horowitz AJ. Sediment source fingerprinting as an aid to catchment management: A review of the current state of knowledge and a methodological decision-tree for end-users. Journal of Environmental Management. 2016;**194**:86-108

[18] Nosrati K. Ascribing soil erosion of hillslope components to river sediment yield. Journal of Environmental Management. 2017;**194**:63-72

[19] Evans GV. Tracer techniques in hydrology. The International Journal of Applied Radiation and Isotopes. 1983;**34**:451-475

[20] Sauzay G. Tracer techniques in sediment transport: Report of the panel. In: Tracer Techniques in Sediment Transport. Technical Report Series No. 145. Vienna, Austria: International Atomic Energy Agency (IAEA); 1973. p. 3-8

[21] Foster IDL. Tracers in Geomorphology. UK: John Wiley & Sons Ltd.; 2000

[22] Heckrath G, Halekoh U, Djurhuus J, Govers G. The effect of tillage direction on soil redistribution by mouldboard ploughing on complex slopes. Soil & Tillage Research. 2006;**88**:225-241

[23] Spokas K, Forcella F, Archer D, Reicosky D. SeedChaser: Vertical soil tillage distribution model. Computers and Electronics in Agriculture. 2007;**57**:62-73

[24] Tiessen KHD, Mehuys GR, Lobb DA, Rees HW. Tillage erosion within potato production systems in Atlantic Canada I. Measurement of tillage translocation by implements used in seedbed preparation. Soil & Tillage Research. 2007;**95**:308-319

[25] Lobb DA, Kachanoski RG. Quantification of tillage translocation and tillage erosion. Canadian Journal of Soil Science. 1994;**74**:353

[26] Liu G, Hai X, Liu P, Zhang J. Using rare earth elements to monitor sediment sources from a miniature model of a small watershed in the three gorges area of China. Catena. 2016a;**142**:114-122

[27] Liu G, Hai X, Liu P, Zhang Q, Zhang J. An improved method for tracing soil erosion using rare earth elements. Journal of Soils and Sediments. 2016b;**16**(5):1670-1679

[28] Zhang XC, Friedrich JM, Nearing MA, Norton LD. Potential use of rare earth oxides as tracers for soil erosion and aggregation studies. Soil Science Society of America Journal. 2001;**65**:1508-1515

[29] Lang A. Recent advances in dating and source tracing of fluvial deposits. In: Sediment Dynamics in Changing Environments. International Association of Hydrological Sciences (IAHS), Publication No. 325. 2008. p. 3-12

[30] Fullen MA. Laboratory and field studies in the use of the isotope ^{59}Fe for tracing soil particle movement. Earth Surface Processes and Landforms. 1982;**7**:285-293

[31] Wooldridge DD. Tracing soil particle movement with Fe-59. Soil Science Society of America Journal. 1965;**29**:469-472

[32] Ventura E, Nearing M, Norton LD. Developing a magnetic tracer to study soil erosion. Catena. 2001;**43**:277-291

[33] Parsons AJ, Wainwright J, Abrahams AD. Tracing sediment movement in interrill overland flow on a semi-arid grassland hillslope using magnetic susceptibility. Earth Surface Processes and Landforms. 1993;**18**:721-732

[34] Plante AF, Duke MJM, McGill WB. A tracer detectable by neutron activation for soil aggregation and translocation studies. Soil Science Society of America Journal. 1999;**63**:1284-1290

[35] Riebe B. Monitoring the translocation of soil particles using a neutron activated tracer. In: Hartge KH, Stewart BA, editors. Soil Structure: Its Development and Function. Florida, USA: CRC Press Inc.; 1995. p. 277-294

[36] Walling DE. Quantifying the fine sediment budgets of river basins. National Hydrology. Seminar. 2004:9-20

[37] Campbell BL, Loughran RJ, Elliott GL. A method for determining sediment budgets using caesium-137. In: Sediment Budgets. International Association of Hydrological Sciences (IAHS), Publication No. 174. 1988. p. 171-179

[38] Greenwood P. Tracing Fine-Sediment Using Artificial Radionuclides. British Society for Geomorphology Publication; 2012 http://britishgeomorphology.org.uk/sites/default/files/geom_tech_chapters/3.5.2_SedimentTracing.pdf

[39] Greenwood P, Walling DE, Quine TA. Assessing the remobilization of recently deposited sediment from river floodplains during single overbank flood events, using caesium-134 and cobalt-60 as tracers. International Association of Hydrological Sciences Publication, Number 325. 2008:13-23

[40] Greenwood P, Walling DE, Quine TA. Using caesium-134 and cobalt-60 as tracers to assess the remobilization of recently-deposited overbank-derived sediment on river floodplains during subsequent inundation events. Earth Surface Processes and Landforms. 2013;**39**(2):228-244

[41] Wallbrink PJ, Olley JM, Hancock G. Estimating residence times of fine sediment in river channels using fallout ^{210}Pb. In: The Structure, Function and Management Implications of Fluvial Sedimentary Systems. International Association of Hydrological Sciences (IAHS), Publication No. 276. 2002a. p. 425-432

[42] Walling DE, Owens PN. The role of flood plain sedimentation in catchment sediment and contaminant budgets. In: The Structure, Function and Management Implications of Fluvial Sedimentary Systems. International Association of Hydrological Sciences (IAHS), Publication No. 276. 2002. p. 407-416

[43] He Q, Walling DE. Interpreting particle size effects in the adsorption of ^{137}Cs and unsupported ^{210}Pb by mineral soils and sediments. Journal of Environmental Radioactivity. 1996;**30**:117-137

[44] Walling DE, He Q. Rates of overbank sedimentation on the floodplains of several British rivers during the past 110 years. In: Variability in Stream Erosion and Sediment Transport. International Association of Hydrological Sciences (IAHS) Publication, No. 224. 1994. p. 77-86

[45] He Q, Walling DE, Wallbrink PJ. Alternative methods and radionuclides for use in soil-erosion and sedimentation investigations. Chpt. 9. In: Zapata F, editor. (2002) Handbook for the Assessment of Soil Erosion and Sedimentation Using Environmental Radionuclides. London: Kluwer Academic Publishers; 2002. p. 185-215

[46] Blake WH, Walling DE, He Q. Fallout beryllium-7 as a tracer in soil erosion investigations. Applied Radiation and Isotopes. 1999;**51**:599-605

[47] Blake WH, Walling DE, He Q. Using cosmogenic Beryllium-7 as a tracer in sediment budget investigations. Geografiska Annaler A. 2002;**84**:89-102

[48] Mabit L, Benmansour M, Walling DE. Comparative advantages and limitations of the fallout radionuclides ^{137}Cs, $^{210}Pb_{ex}$ and ^{7}Be for assessing soil erosion and sedimentation. Journal of Environmental Radioactivity. 2008;**99**:1799-1807

[49] Walling DE, Schuller P, Zhang Y, Iroumé A. Extending the timescale for using beryllium-7 measurements to document soil redistribution by erosion. Water Resources Research. 2009;**45**:W02418

[50] McCubbin D, Leonard KS. Use of radiotracers for studies of metal sorption behaviour. Science of the Total Environment. 1995;**173/174**:259-266

[51] Showler AT, Knaus RM, Reagan TE. The versatility of radiotracer methods for studying insect ethology and ecology. Florida Entomologist. 1988;**71**(4):554-580

[52] Kinnell PIA. The effect of flow depth on sediment transport induced by raindrops impacting shallow flows. America Society of Agricultural Engineers. 1991;**34**(1):161-168

[53] Quine TA, Govers G, Poesen J, Walling DE, van Wesemael B, Martinez-Fernandez J. Fine-earth translocation by tillage in stony soils in the Guadalentin, south-east Spain: An investigation using caesium-134. Soil & Tillage Research. 1999;**51**:279-301

[54] Syversen N, Øygarden L, Salbu B. Cesium-134 as a tracer to study particle transport processes within a small catchment with a buffer zone. Journal of Environmental Quality. 2001;**30**:1771-1783

[55] Toth SJ, Alderfer RB. A procedure for tagging water-stable aggregates with Co-60. Soil Science. 1959a;**89**:36-37

[56] Toth SJ, Alderfer RB. Formation and breakdown of Co-60-tagged water-stable aggregates in a Norton Silt Loam Soil. Soil Science. 1959b;**90**:232-238

[57] Ridley N. The Radioactive Substances (Substances of Low Activity) Exemption Order 1986. Statutory Instruments, No 1002. Atomic Energy and Radioactivity Substances. 1986. http://www.ionactive.co.uk/pdfs/Substances_Low_Activity_Exemption_Order_11190 18891.pdf [Accessed 25/07/17]

[58] Greenwood P, Walling DE, Quine TA. Accepted. Documenting sediment redistribution on livestock-poached pasture using the artificial radionuclides caesium-134 and cobalt-60 as tracers. Land degradation and development. Special Issue: The impact of grazing on land degradation: A common problem with many driving forces

[59] Khan SA. Sorption of the long-lived radionuclides cesium-134, strontium-85 and cobalt-60 on bentonite. Journal of Radioanalytical and Nuclear Chemistry. 2003;**258**:3-6

[60] Environmental Protection Agency (EPA). Radiation Protection: Cobalt. USA: Environmental Protection Agency; 2002 www.epa.gov/rpdweb00/radionuclides/cobalt.html [Accessed 25/07/17]

[61] Bailly du Bois P. Mapping of water masses in the North Sea using radioactive tracers. Endeavour. 1996;**20**:2-7

[62] Cundy AB, Croudace IW. Sediment accretion and recent sea-level rise in the Solent, Southern England: Inferences from radiometric and geochemical studies. Estuarine, Coastal and Shelf Science. 1996;**43**:449-467

[63] Matishov GG, Matishov DG, Namjatov AA, Carroll JL, Dahle S. Anthropogenic radionuclides in Kola and Motovsky Bays of the Barents Sea, Russia. Journal of Environmental Radioactivity. 1999;**43**:77-88

[64] Bailly du Bois P, Guéguéniat P. Quantitative assessment of dissolved radiotracers in the English Channel: Sources, average impact of la Hague reprocessing plant and conservative behaviour (1983, 1986, 1988, 1994). Continental Shelf Research. 1999;**19**:1977-2002

[65] Thompson J, Dyer FM, Croudace IM. Records of radionuclide deposition in two salt marshes in the United Kingdom with contrasting redox and accumulation conditions. Geochimica et Cosmochimica Acta. 2002;**66**(6):1011-1023

[66] Sutherland RA. Spatial variability of ^{137}Cs and the influence of sampling on estimates of sediment redistribution. Catena. 1994;**21**:57-71

[67] Caron F, Mankarios G. Pre-assessment of the speciation of ^{60}Co, ^{125}Sb, ^{137}Cs and ^{241}Am in a contaminated aquifer. Journal of Environmental Radioactivity. 2004;**77**.29-46

[68] Payne TE, Itakura T, Comarmond MJ, Harrison JJ. Environmental mobility of cobalt – Influence of solid phase characteristics and groundwater chemistry. Applied Radiation and Isotopes. 2009;**67**:1269-1276

[69] Shinonaga T, Schimmack W, Gerzabek MH. Vertical migration of ^{60}Co, ^{137}Cs and ^{226}Ra in agricultural soils as observed in lysimeters under crop rotation. Journal of Environmental Radioactivity. 2005;**79**:93-106

[70] Environmental Protection Agency (EPA). Understanding Variation in Partition Coefficient K_d Values – Attachment C: Radiological Properties for SSL Development. USA: Office of Air & Radiation, Environmental Protection Agency; 1999

[71] Chen L, Lu S. Sorption and desorption of radiocobalt on montmorillonite—Effects of pH, ionic strength and fulvic acid. Applied Radiation and Isotopes. 2008;**66**:288-294

[72] Sparks DL. Environmental Soil Chemistry. 2nd ed. Amsterdam, The Netherlands: Academic Press, Elsevier Publishers Ltd.; 2003

[73] Shor JT, Dial C. Volatilization of cobalt, technetium, and uranium isotopes from soils through salt amendment and calcination. Journal of Environmental Radioactivity. 2000;**48**:35-48

[74] Capowiez Y, Renault P, Belzunces L. Three-dimensional trajectories of ^{60}Co-labelled earthworms in artificial cores of soil. European Journal of Soil Science. 2001;**52**:365-375

[75] Sattar A, Traniello JFA, Salihah Z, Farid A. Determining different parameters of Co-60 bait as a radiotracer for three species of subterranean termites. The. Nucleus. 2002;**39**:89-93

[76] Wallbrink PJ, Walling DE, He Q. Radionuclide measurement using HPGe gamma spectrometry. Cp. 5. In: Zapata F, editor. (2002) Handbook for the Assessment of Soil Erosion and Sedimentation Using Environmental Radionuclides. London: Kluwer Academic Publishers; 2002b. p. 67-96

[77] Pennock DJ, Appleby PG. Site selection and sampling design, Chpt. 2. In: Zapata F, editor. Handbook for the Assessment of Soil Erosion and Sedimentation Using Environmental Radionuclides. London, UK: Kluwer Academic Publishers; 2002. p. 15-40

[78] Owens PN, Walling DE, Leeks GJL. Deposition and storage of fine-grained sediment within the main channel system of the River Tweed, Scotland. Earth Surface Processes and Landforms. 1999;**24**:1061-1076

[79] Walling DE, Owens PN, Carter J, Leeks GJL, Lewis S, Meharg AA, Wright J. Storage of sediment-associated nutrients and contaminants in river channel and floodplain systems. Applied Geochemistry. 2003;**18**:195-220

[80] Nicholas AP, Walling DE, Sweet RJ, Fang X. New strategies for upscaling high-resolution flow and overbank sedimentation models to quantify floodplain sediment storage at the catchment scale. Journal of Hydrology. 2006;**329**:577-594

[81] Simm DJ, Walling DE. Lateral variability of overbank sedimentation on a Devon floodplain. Hydrological Sciences. 1998;**43**:715-732

[82] Le Bayon RC, Binet F. Rainfall effects on erosion of earthworm casts and phosphorous transfers by water runoff. Biological Fertility of Soils. 1999;**30**:7-13

[83] Le Bayon RC, Binet F. Earthworm surface casts affect soil erosion by run-off and phosphorous transfer in a temperate maize crop. Pedobiologia. 2001;**45**:430-442

[84] Le Bayon RC, Moreau S, Gascuel-Odoux C, Binet F. Annual variations in earthworm surface-casting activity and soil transport by water run-off under a temperate maize agroecosystem. Geoderma. 2002;**106**:121-135

[85] Collins AL, Walling DE, Leeks GJL. Source type ascription for fluvial suspended sediment based on a quantitative composite fingerprinting technique. Catena. 1997;**29**:1-27

[86] Bilotta GS, Brazier RE, Haygarth PM, MacLeod CJA, Butler P, Granger S, Krueger T, Freer J, Quinton J. Rethinking the contribution of drained and undrained grasslands to sediment-related water quality problems. Journal of Environmental Quality. 2008;**37**:906-914

[87] Orr RJ, Murray PJ, Eyles CJ, Blackwell MSA, Cardenas LM, Collins AJ, Dungait JAJ, Goulding KWT, Griffith BA, Gurr SJ, Harris P, Hawkins JMB, Misselbrook TH, Rawlings C, Shepard A, Sint H, Takahashi T, Tozer KN, Whitmore AP, Wu L, Lee MRF. The North Wyke Farm Platform: Effect of temperate grassland farming systems on soil moisture contents, runoff and associated water quality dynamics. European Journal of Soil Science. 2016;**67**:374-385

[88] Granger SJ, Bol R, Butler PJ, Haygarth PM, Naden P, Old G, Owens PN, Smith BPG. Processes affecting transfer of sediment and colloids, with associated phosphorous, from intensively farmed grasslands: Tracing and organic matter. Hydrological Processes. 2007;**21**:417-422

[89] Walling DE, He Q, Appleby PG. Conversion models for use in soil-erosion, soil-redistribution and sedimentation investigations. Chpt. 7. In: Handbook for the Assessment of Soil Erosion and Sedimentation Using Environmental Radionuclides, F Zapata. London, UK: Kluwer Academic Publishers; 2002. p. 111-164

[90] Alam SM, Ansari R, Khan A. Application of radioisotopes and radiation in the field of agriculture: Review. Journal of Biological Sciences. 2001;**1**(3):82-86

[91] Oldeman LR. Soil Degradation: A Threat to Food Security? Report 98/01. International Soil Reference and Information Centre: Wageningen; 1998

[92] Ritchie JC, McHenry JR. Application of radioactive fallout cesium-137 for measuring soil erosion and sediment accumulation rates and patterns: A review. Journal of Environmental Quality. 1990;**19**:215-233

[93] Schreiner LJ, Joshi CP, Darko J, Kerr A, Salomons G, Dhanesar S. The role of cobalt-60 in modern radiation therapy: Dose, delivery and image guidance. Journal of Medical Physics. 2009;**34**(3):133-136

[94] Thorén M, Rahn T, Guo WY, Werner S. Stereotactic radiosurgery with the cobalt-60 gamma unit in the treatment of growth hormone-producing pituitary tumors. Neurosurgery. 1991;**29**:663-668